SpringerBriefs in Molecular Science

More information about this series at http://www.springer.com/series/8898

Gábor Lente

Deterministic Kinetics in Chemistry and Systems Biology

The Dynamics of Complex Reaction Networks

 Springer

Gábor Lente
Department of Inorganic and Analytical Chem.
University of Debrecen
Debrecen, Hungary

ISSN 2191-5407   ISSN 2191-5415   (electronic)
SpringerBriefs in Molecular Science
ISBN 978-3-319-15481-7   ISBN 978-3-319-15482-4   (eBook)
DOI 10.1007/978-3-319-15482-4

Library of Congress Control Number: 2015932083

Springer Cham Heidelberg New York Dordrecht London
© Gábor Lente 2015

Printed on acid-free paper

Springer International Publishing AG Switzerland is part of Springer Science+Business Media
(www.springer.com)

Preface

Story has that Henry Taube, shortly before winning the 1983 Nobel Prize in chemistry, considered his life's work a failure. The reason why he felt so was that he did not write a textbook on inorganic chemistry. Instead, Al Cotton's textbook became widely popular, whose text had the main emphasis on structure instead of reactivity, which Taube did not like at all. Even if the story is not true, its message is crystal clear: writing a textbook gives the author a chance to influence the scientific thinking of future generations.

Another Nobel Laureate, Roald Hoffmann, wrote quite eloquently about some scientists' "desire to convince, to scream, 'I'm right, all of you are wrong', clashing with the established rules of civility supposedly governing scholarly behavior" in his book titled *The same and not the same*. This burning feeling is indeed very familiar to the present author, who encounters scientific lines of thought that he believes to be incorrect daily.

This book is my brief account of chemical kinetics. It mainly presents how kinetic curves should be evaluated and how kinetic experiments can be designed to maximize their information content. There should only be one good reason to write a book: to say something that has never been said before. Yet, textbooks on chemical kinetics are available in a considerable variety. Why do I think my book is unique then?

I hope the reader did not expect a short answer to the previous question. I was fortunate enough to reinvent (established rules of scholarly civility oblige me to use this word, but the actual feeling at the time always was the excitement of invention) many ways to solve kinetic problems. Some of these problems are quite common; others are curiosities even for experts. After some time, a few of my colleagues took notice of this fact and began asking occasional questions, most of which I could answer. This process has been on for a long enough time now to convince me that this expertise may help other scientists in their work. Therefore, in this book, I tried to summarize these problem-solving strategies in a systematic way. The text introduces the basic concepts of chemical kinetics but typically also contains my personal opinion as well. The reader will probably be surprised by some of the remarks as I did not refrain from criticizing kinetic techniques that I consider wrong despite the fact that they have a long history and would probably qualify as "accepted practice" for many scientists. Logic tells me that mathematical derivations or proofs are not a matter of opinion or acceptance, they are either right or wrong (although any person can make a mistake in judgment). My personal opinion, on the other hand, may and hopefully, will be debated.

I intended this book to be a bridge between theoreticians and experimentalists. For theoreticians, the text tries to present the practical significance of concepts and mathematical techniques, always explaining how the assumptions made relate to physical reality. On the other hand, I felt that it would be useful for experimentalists to have a much wider picture of mathematical possibilities than that available in their commonly used textbooks.

I kept the number of literature references intentionally low in this book. This is intended as a service to the reader. In our Internet age, searching in the scientific literature has become very easy. A much more difficult task is assessing the search results in terms of reliability and significance. My objective was to write a book that can be understood and used without checking any of the previous literature. References usually serve one of two purposes (sometimes both): (1) To pinpoint a source that provides mathematical proofs or other background information, which I do not see as vital, but may be useful for an advanced reader. (2) To acknowledge the priority of a scientist who is widely believed to introduce a certain concept.

All the derivations and mathematical lines of thought presented in this text have been meticulously repeated by the author; nothing was simply taken from the literature without rethinking and cross-checking. The result is that the author is responsible for any possible mistakes in this book: hopefully, these are typos only and not failures of logical thinking.

This book tries to present how kinetic measurements are best done and evaluated in modern research. It could be ironic that the author, or indeed any living person, feels entitled to undertake such a task. The only excuse for this gross immodesty can be that scientists (young or more experienced alike) might benefit from reading the results of this effort—the book.

Debrecen, Hungary   Gábor Lente
November 2014

Acknowledgements

I express my heartfelt gratitude to István Fábián (Debrecen, Hungary), Jim Espenson (Ames, IA, USA), Péter Érdi (Kalamazoo, MI, USA), István Bányai (Debrecen, Hungary), and János Tóth (Budapest, Hungary).

István Fábián was my mentor and teacher during my years as an undergraduate and graduate student. Our collaboration has a 20-year-old history now, and we still work in the same research group at the University of Debrecen. In addition to his scientific guidance, I am also grateful to him for creating opportunities to follow my own research interests.

I first knew Jim Espenson as the author of a textbook on chemical kinetics and reaction mechanisms. I still value this book very highly and recommend it to everyone. I also had the good fortune to work with him at Iowa State University in Ames for 2 years and try to learn from him directly. I wish him very happy years in retirement.

Péter Érdi undertook the burden of coauthoring a book with me on stochastic kinetics. Writing this earlier book convinced me that I actually have a lot to say about deterministic kinetics as well. Péter never denied me help from his immense experience, which he typically shares within 10 min of receiving an e-mail from me.

István Bányai is a very open-minded chemist and János Tóth is a very open-minded mathematician. Neither of them can say no to a highly stimulating conversation about any topic, and least of all, chemical kinetics.

Writing of this book was supported by the EU and co-financed by the European Social Fund under the project ENVIKUT (TÁMOP-4.2.2.A-11/1/KONV-2012-0043).

Last, but not least, I also thank my wife Kata, who is a constant source of inspiration. In addition, as a professor of physical chemistry, she does not only listen to my ideas but also comments on them regularly.

Contents

Chapter 1
Rates and Rate Equations

In general, deterministic kinetics is suitable to describe the time evolution of various physical quantities. The most elaborated use of kinetics is probably in studying chemical reactions, but other, quite diverse scientific disciplines use the same principles. For example, population dynamics in biology may use very similar equations to describe temporal changes in the number of individual living beings. Recently, systems biology is being developed and essentially does the same as conventional chemical kinetics—just in a biochemical context. Even if the purposes and the physical quantities used are quite different, the underlying mathematics is the same in those cases, which means that studying these principles in one discipline often results in knowledge that is directly transferable to another.

1.1 Concentration and Its Change in Time

The size of a group can be quantified in different ways. In chemistry, it is possible to count the number of individual entities (molecules, molecule fragments, radicals, ions, atoms). The concept of **amount of substance** is one of the fundamental properties in the widely used Système International d'Unités (SI), and is measured in moles (1 mol roughly equals 6×10^{23} entities). The size of the investigated system is typically irrelevant in describing changes in time. Therefore, an intensive quantity, **concentration** (amount of substance divided by the volume of the system) is preferred. Its SI unit would be mol/m^3, but the most common unit in use is mol/dm^3, which has a common one-letter abbreviation (M). For measuring concentration, mmol/dm^3 (which is the same as mol/m^3), μmol/dm^3, or even nmol/dm^3 is also used in certain cases. In gas kinetics, the concentration unit molecule/cm^3 (1.66×10^{-21} mol/dm^3) is also very common, and to make matters confusing, this unit is sometimes abbreviated as mol./cm^3. The dot is often lost

© Gábor Lente 2015
G. Lente, *Deterministic Kinetics in Chemistry and Systems Biology*,
SpringerBriefs in Molecular Science, DOI 10.1007/978-3-319-15482-4_1

during typesetting, and this mistake creates a high potential of confusing mol./cm^3 with mol/cm^3 ($= 10^6$ mol/dm^3). The numerical values of concentrations will give such a mistake away.

In real-life cases, there is typically more than just one kind of entities that are of interest. Let $A_1, A_2, \ldots$ be different entities present in the system. In chemistry, these entities are particles of matter, molecules, or ions. The concentrations are symbolized by $[A_1], [A_2], \ldots$, so the brackets around the symbols of the entities are commonly used for this purpose. Another common notation would be $c_{A_1}, c_{A_2}, \ldots$, but the need for multiple subscripts makes this notation somewhat less practical.

Kinetics seeks the concentration of entities as a function of time. Therefore, $[A_1], [A_2], \ldots$ do not simply denote values, but functions that have time as an independent variable. Confusing functions with their values taken at particular time instances is a major source of error in many sequences of thought. Time zero is usually naturally selected by the experiment (e.g., the time instant of mixing). When it is necessary, the **initial concentration** (at $t = 0$) of component A_i will be denoted $[A_i]_0$, the value at time $t = \tau$ is $[A_i]_\tau$, and the **final concentration** (at $t = \infty$), which is not guaranteed to exist (because, e.g., infinite periodic change of a concentration is not impossible), will be denoted $[A_i]_\infty$. Distinguishing between a particular time instant from the general notion of time (t) is seldom highly important, but failure to do so is sometimes a source of erroneous derivations.

In chemistry, experiments are carried out in a container called **reactor**, which typically has a constant volume and is **closed**, i.e., no particles are exchanged between the reactor and its environment. The chemical entities are quantized, so concentration should also be quantized: it assumes only certain discrete values that are integer multiples of a smallest possible concentration, which is 1 entity/reactor and its value in M depends on the overall volume of the reactor. However, the numbers of entities are typically so high that approximating concentration as a mathematically continuous variable is quite acceptable. This is an important point: using continuous concentrations greatly simplifies the mathematics of describing the temporal changes. This simplifying assumption is central in deterministic kinetics. If it is not valid, e.g., the reactor contains a small overall number of entities, deterministic kinetics should be replaced by the approach of stochastic kinetics [2].

The rate of concentration change is best described by the derivative of the concentration with respect to time, which is the change in concentration divided by the change in time when the time interval tends to infinitely low values. The derivative is also a function of time and not a single value. In a mathematical form, the definition of the rate of concentration change for entity A_i can be given as follows:

$$\left(\frac{d[A_i]}{dt}\right)_\tau = \lim_{\delta \to 0} \frac{[A_i]_{\tau+\delta} - [A_i]_\tau}{\delta} \tag{1.1}$$

This equation defines the derivative at time instant τ. In one of the previous paragraphs, it was emphasized that concentration is assumed to be a continuous function of time. The use of the derivative introduces a still stricter condition: concentration is a differentiable function of time.[1]

The differential equation commonly referred to as the **rate equation** gives the rates of concentration change as a function of the concentrations of n different entities present in the system. In this text, the following representation will be used:

$$\frac{d[A_i]}{dt} = f_i([A_1], [A_2], \ldots, [A_n]) \tag{1.2}$$

Functions f_i here are specific to the component A_i. Therefore, the full rate equation is a collection of n different functions, all of which may have all the concentrations as independent variables. Furthermore, f_i functions are typically continuous, though examples to the contrary exist, which always need extra attention.

This text defines the rate equation well before (and, indeed, instead of) the **rate of reaction**, which may be uncommon in textbooks about chemical kinetics, but does not lead to contradictions. Rates of concentration change can be interpreted without defining the reaction rate, which is an ambiguous concept. The rate equation is thought to describe the rates of concentration change in this book and not the reaction rate. In a few cases, a quantity called the advancement of reaction, abbreviated as ξ, is also defined by dividing concentration changes with stoichiometric coefficients and the reaction rate is given as the derivative of the advancement of reaction. This is not seen is a productive line of thought in this book.[2]

Rate equation (1.2) assumes a **homogeneous** system, which means that the intensive physical properties within the reactor (most importantly, the concentrations) do not depend on the spatial coordinates. The description of such a system is independent of the values of extensive physical properties, including the volume of the reactor. As a rule, homogeneity is a self-conserving property, i.e., an initially homogeneous system will conserve homogeneity unless it is under a special, direction-dependent external influence. In experiments, it is sufficient to stir the system initially. When homogeneity is reached, stirring is no longer necessary. In non-homogeneous systems, the concentrations of the entities are also a function

---

[1] For those interested in precise mathematics: concentration is indeed a continuous function of time, but it can be non-differentiable in isolated points. Some examples of non-differentiable points on concentration-time curves will be given in Chap. 2.

[2] William of Ockham, an English Franciscan friar, scholastic philosopher and theologian in the fourteenth century famously introduced a logical guideline for science, which is most often quoted today with the sentence "Entia non sunt multiplicanda praeter necessitatem" (Entities must not be multiplied beyond necessity) despite the fact that this actual phrasing never appears in Occam's works known today. The principle is known as Occam's razor. The present author views the concepts of the reaction rate and the advancement of reaction as unnecessary in chemical kinetics. Therefore, Occam's razor calls for avoiding their use.

of spatial coordinates and partial differential equations also allowing for diffusion are necessary to describe their change. These systems have considerable practical importance, but a separate book would be needed to deal with them.

Another remark should be made here about the term **deterministic kinetics**. If the initial concentrations are known with certainty, then the values of the concentrations are unambiguously determined for the entire process by Eq. (1.2). In contrast, stochastic kinetics can only give probabilistic information of the time evolution even if the initial conditions are known for certain [2].

Typically, the volume of the system does not change during an experiment. When this is not true, the left-hand side of the rate equation (1.2) needs to be modified to account for the concentration change occurring without a change in the number of entities. This can be achieved as follows:

$$\frac{d[A_i]}{dt} + \frac{[A_i]}{V}\frac{dV}{dt} = f_i([A_1], [A_2], \ldots, [A_n]) \tag{1.3}$$

The volume change as the function of time in such cases may be known from independent sources (e.g., flow rates). Alternatively, the volume could be determined by the concentration, which might seem paradoxical at first. A theoretically interesting case would be an **isobaric gas reactor**, in which the volume can change so that the pressure remains constant. The concept may be unusual in kinetics, but is directly analogous to the thermodynamics of processes at constant pressure.

As kinetics is concerned with rate equations, a kinetic study necessarily involves measurements of rates of concentration change. Unfortunately, there are no convenient experimental techniques that measure the rates directly. Concentrations, on the other hand, can be determined by a huge variety of methods. Rates are typically determined by monitoring the concentration as a function of time and then using numerical differentiation. Applied mathematics warns that numerical differentiation typically implies large uncertainties: determining rates of concentration change is not nearly as precise as determining concentrations. In addition, determining concentrations is not necessarily needed for a kinetic study. It is often enough to rely on an instrumental signal that is proportional to the concentrations without actually calculating the values.

Simple physical constraints seriously limit the mathematical form of the functions f_i in the rate equation. An obvious limitation arising from the non-negativity of entity numbers is that the rate of concentration change of component A_i cannot be negative when it is absent (i.e., $[A_i] = 0$). Yet in practice, rate equations violating this trivial criterion are also used (a common example is the zeroth order rate equation). The present author calls such rate equations **incomplete** as they cannot be used for concentration values close to zero. Not recognizing the incomplete nature of a rate equation posits a minor, but nonetheless very real danger to logical thinking. A similar phenomenon may occur when one or more of the functions f_i are not interpreted for certain combinations of concentrations that are otherwise possible. A more nuanced case is when f_i functions seem to be regular (bounded and non-negative at zero concentrations), but give a differential equation that leads to a point of singularity at a finite time value. Most of these cases are nice little oddities that seldom have any practical importance.

1.2 Reactions and the Reducibility of Chemical Systems

The primary goal of kinetics is to explore the functions f_i in the rate equation and understand their origins, which means the molecular background of the functions. Understanding is thought to be achieved if the concentration change is interpreted by a chemical reaction or a sequence of reactions.

An overall chemical reaction is thought to consist of a finite number of individual reaction steps. Chemical reaction steps possible in the system are represented by stoichiometric equations, which have the following form for a reaction system that features n different species $(A_1, A_2, \ldots, A_n)$ of interest:

$$0 = \sum_{i=1}^{n} v_{j,i} A_i \quad (j = 1, 2, \ldots, m) \tag{1.4}$$

In Eq. (1.4), m is the number of different reactions. The value $v_{j,i}$, called the stoichiometric coefficient of component A_i in reaction step j, is positive for species that are produced (**products**), negative for species that are consumed (**reactants**), and 0 for species that do not appear in reaction step j. Traditionally, only integers are used as stoichiometric coefficients, with 1 as their greatest common factor in any reaction (a given value of j in Eq. (1.4)). This is especially important in mass action type kinetics (see later in this section). The matrix composed of the stoichiometric coefficients is called the **stoichiometric matrix** of the system:

$$\underline{\underline{v}} = \begin{pmatrix} v_{1,1} & v_{1,2} & \cdots & v_{1,n} \\ v_{2,1} & v_{2,2} & \cdots & v_{2,n} \\ \vdots & \vdots & \ddots & \vdots \\ v_{m,1} & v_{m,2} & \cdots & v_{m,n} \end{pmatrix} \tag{1.5}$$

It is typically possible to show that certain linear combinations of the concentrations are constant (i.e., do not depend on time, only on the initial conditions). These are referred to as conservation equations, which arise from well-known physical laws such as conservation of matter and conservation of charge in closed systems. As chemical processes only rearrange the atoms but do not change them, there is one conservation equation for each type of atom that appears in the process and one for electrical charge, although some of them are not necessarily independent.

It is common to represent stoichiometric equations using a reaction arrow. On the left, reactants are shown, whereas products are displayed on the right. For example, the stoichiometric equation $0 = -A_1 - 3A_2 + 2A_3$ is given as $A_1 + 3A_2 \longrightarrow 2A_3$ in this formalism.

Despite the fact that textbooks define the concept of the general rate of reaction by dividing the rate of concentration change of a particular entity by the stoichiometric coefficient, this definition should be limited to single-step reactions. IUPAC[3] recommendations also point out this restriction [5, 8]. In other words, the rate of reaction should generally be a vector whose dimension is determined by the number of reactions steps. The rate of reaction step j (v_j) can be defined if necessary and the f_i functions can be given as the sum of the rates of the individual steps:

$$f_i([A_1], [A_2], \dots, [A_n]) = \sum_{j=1}^{m} v_{j,i} v_j \quad (i = 1, 2, \dots, n) \tag{1.6}$$

In addition to concentrations, the rates of an individual reaction steps (v_j) depend on some parameters usually called rate constants and denoted as k_j. Typically, different rates have different such parameters, but symmetry laws may result in different reaction steps having identical rate constants.

For a single-step reaction ($j = 1$), the rate of reaction is defined as:

$$v = \frac{1}{v_{1,i}} \frac{d[A_i]}{dt} \tag{1.7}$$

Many experimentally encountered systems obey **power law kinetics**, which means that rates v_j can be obtained by multiplying the concentrations raised to a suitable power as shown in the following equation:

$$v_j = k_j \prod_{k=1}^{n} [A_k]^{\alpha_{j,k}} \tag{1.8}$$

In this case, the f_i functions are given as follows:

$$f_i([A_1], [A_2], \dots, [A_n]) = \sum_{j=1}^{m} v_{j,i} k_j \prod_{k=1}^{n} [A_k]^{\alpha_{j,k}} \quad (i = 1, 2, \dots, n) \tag{1.9}$$

The $\alpha_{j,k}$ values are often, but not necessarily, integer and are called the **order of reaction** for step j with respect to substance A_k. The sum $\sum_{k=1}^{n} \alpha_{j,k}$ is called the overall order of reaction step j, which has special significance in determining

---

[3]IUPAC stands for the International Union of Pure and Applied Chemistry, an organization of chemist that issues all sorts of recommendations and oversees chemical nomenclature.

the physical dimension of the rate constant k_j. In general, values of $\alpha_{j,k}$ cannot be deduced from stoichiometric coefficients and can be collected in an **order matrix**:

$$\underline{\underline{\alpha}} = \begin{pmatrix} \alpha_{1,1} & \alpha_{1,2} & \cdots & \alpha_{1,n} \\ \alpha_{2,1} & \alpha_{2,2} & \cdots & \alpha_{2,n} \\ \vdots & \vdots & \ddots & \vdots \\ \alpha_{m,1} & \alpha_{m,2} & \cdots & \alpha_{m,n} \end{pmatrix} \qquad (1.10)$$

The dependence of rate constants k_j on external conditions (most prominently on temperature) can usually be neglected during a single experiment, so these parameters are constants as far as the solution of Eq. (1.2) is concerned. Should this not be the case, the notation of power law kinetics is usually retained and additional differential equations are introduced to describe the time dependence of the parameters. A system with nonconstant volume can also be handled with this technique. A special difficulty arises here because concentrations change not only in chemical reactions, but also as a result of the volume change (see Eq. (1.3)).

Mass action type kinetics is a special case of power law kinetics and is characterized by the fact that the order matrix can be determined from the stoichiometric matrix using the following simple rule:

$$\alpha_{j,i} = -\nu_{j,i} \quad \text{if} \quad \nu_{j,i} < 0$$

$$\alpha_{j,i} = 0 \quad \text{if} \quad \nu_{j,i} \geq 0 \qquad (1.11)$$

It is a very basic (but, most inappropriately, typically only implicit) postulate of chemical kinetics that observations in all systems can be described by special mass action type kinetics called a series of **elementary reactions**. This will be referred to as the postulate of **reducibility** in this book. Unfortunately, the concept of an elementary reaction is not very clearly defined in a mathematical sense. The IUPAC recommendations give the definition as "A reaction for which there is no evidence that it occurs in more than one step is assumed to occur in one step and is said to be an elementary reaction [8]." or "A reaction for which no reaction intermediates have been detected or need to be postulated in order to describe the chemical reaction on a molecular scale [5]."

From a purely logical point of view, these are highly insufficient ways of defining the (otherwise very important) concept of the elementary reaction as they often lead to circular arguments, and say nothing about the typical properties. Most kineticists seem to take a somewhat intuition-based approach to elementary reactions and recognize them by their properties. As stated, one of the final goals of chemical kinetics is to reduce an overall reaction into a series of elementary reactions, i.e., to interpret macroscopic observations (concentration changes) on a particle-based level (as a series of simple molecular transformations).

A necessary but not sufficient condition for a mass action type system to qualify as a series of elementary reactions is that all stoichiometric coefficients are 0, ± 1, or ± 2 (possibly ± 3 in very exceptional cases) and none of the reaction steps have an overall reaction order greater than 2 (3 is possible again as a rare exception). An elementary step can only be one of three types. The first is the **unimolecular** elementary step, which involves the transformation of a single particle without interference from other species and is first order with respect to the single molecule involved. The second type is called a **bimolecular** elementary step. It is a reaction between two (identical or different) entities and is either second order to its only reagent or first order with respect to both of its two different reagents. A third, and very rare possibility is a **termolecular** step, in which three molecules are involved in a third order reaction. There is no limitation for the number of products in the three possible types of elementary reactions. A series of elementary reactions gives rise to a rate equation in which only first, second, and third order reactions may occur.

If observations are described by a rate law that cannot represent a series of elementary reactions, this fact usually implies that not all components or reactions have been correctly identified, and a more complete description is possible by taking into account more elementary reactions. However, experimental data often do not allow the identification of missing steps. In these cases, it is entirely up to the judgment of the experimenter to decide whether finding a suitable series of elementary reactions is necessary for research purposes or the simpler, but theoretically incomplete description serves the objectives better.

A common notation used to condense the stoichiometric and kinetic information of a reaction step with a power rate law is to write the following chemical equation:

$$\sum_{i=1}^{n} \alpha_{j,i} A_i \longrightarrow \sum_{i=1}^{n} (\alpha_{j,i} + v_{j,i}) A_i \tag{1.12}$$

In a mathematical sense, this sort of notation is limited to cases when all $\alpha_{j,i}$ and $(\alpha_{j,i} + v_{j,i})$ are non-negative. This is not much of a limitation from a practical point of view as most known processes (e.g., all reactions with mass action type kinetics) satisfy this criterion. Because of its brevity, this notation is more popular than giving separate stoichiometric and order matrices. All series of elementary reactions can be represented unambiguously by this notation.

Further limitations from physical and chemical laws apply to rate equations. Because of the property of reducibility and the difficulty in defining elementary reactions, the mathematical consequences of these limitations are most practically stated in terms of power law kinetics. One obvious limitation, already mentioned previously, is that concentrations should remain non-negative at any reaction time. A sufficient but not necessary condition for the non-negativity of concentrations is that $\alpha_{j,i} > 0$ should hold for any pair of (i, j) values for which $v_{j,i} < 0$. Mass action kinetics not only satisfies this necessary condition, but also guarantees that all component concentrations remain positive (cannot be zero at finite times).

Another set of limitations are imposed by the law of mass conservation. These can often be deduced from the stoichiometric equations in closed systems (i.e., those which cannot exchange particles with the surroundings) and they are also very useful for eliminating some of the concentrations during the solution of the rate equation.

A more special set of limitations is given by the **principle of detailed balance**, which posits that for each stoichiometric reaction (1.4), the model must also contain the exact reverse reaction as well (**microscopic reversibility**):

$$0 = \sum_{i=1}^{n} -v_{r(j),i} A_i \tag{1.13}$$

In Eq. (1.13), $r(j)$ is a function giving the number of the stoichiometric equation corresponding to the reverse of stoichiometric equation j. Furthermore, the principle of detailed balance also requires a relationship between $\alpha_{j,i}$ and $\alpha_{r(j),i}$ values:

$$v_{j,i} = \alpha_{r(j),i} - \alpha_{j,i} \tag{1.14}$$

The ratio of the rate constants of the forward and reverse steps, $k_j / k_{r(j)}$, should be equal to the equilibrium constant of the process, which can be checked against values from independent measurements. However, the required values of reverse rate constants are often so low so that they have no experimentally detectable consequences. Therefore, it is very common to deal with rate laws that violate the principle of detailed balance.

Different types of physical limitations apply to the values of rate constants of power law type rate equations. All rate constants have a lower limit of 0. Reaction steps with exactly 0 rate constants can be deleted from the system. The upper limits of the rate constant values are set either by the time scale of intramolecular motion or the velocity molecules move relative to each other, depending on the overall order of the reaction step (see Sect. 4.1).

For single step reactions, the order of reaction can be defined based on the concentration dependence of the reaction rate:

$$\alpha_{1,i} = \frac{\partial v_1}{\partial [A]_i} \tag{1.15}$$

The partial differentiation is necessary in this formula to emphasize the fact that all other concentrations must be held constant. The order of reaction calculated by this definition may be dependent on the very concentration it refers to ($[A]_i$) or even other concentrations. Although there may be some rationale in using concentration-dependent reaction orders for certain purposes, generally, this is a clear sign showing that the reaction follows a non-power law type rate equation. In cases like these, other functional forms of the rate equation must be considered.

One more important physical postulate, called the **principle of independent interactions**, should also be recalled here. For chemical kinetics, the principle states that a rate constant of an elementary process cannot be influenced by the presence or absence of substances not appearing in it. Therefore, additives that influence an overall reaction (catalyst, inhibitors) can only exert their effect by engaging one or more of the reactants in new elementary reactions.

1.3   Intermediates and Catalysts

The entities in a system can be classified into different types. These classifications are not always unambiguous (or even necessary), but often provide help in understanding changes. Reactants and products have also been mentioned briefly. A **reactant** (or **reagent**) is a species that is initially present and some (or all) of it is consumed in the process: the final concentration is lower than the initial. A **product** is not typically present at the beginning, but appears and accumulates in the system during the course of the studied changes. A reactant that is consumed entirely (so that its final concentration falls practically to zero) is often called a **limiting reagent**. A reactant that is not consumed entirely is called the **excess reagent**. Reactants and products can normally be identified based on the nature of the chemical reaction without quantitative information. The designation of limiting and excess reagents, on the other hand, only makes sense for specified initial concentrations. These roles may even be exchanged under different initial conditions.

Classically, intermediates are substances that are not present neither at the beginning of the process nor at the end. Species A_i is an intermediate if it has nonzero concentrations at some time instances, but the initial and final value is zero:

$$[A_i]_0 = \lim_{\tau \to \infty} [A_i]_\tau = 0 \tag{1.16}$$

An intermediate forms as a result of the reaction steps occurring in the system, then it is consumed in different reaction steps. It is often important to distinguish between major intermediates and minor intermediates. The concentration of a major intermediate rises to values comparable to the initial concentration of the limiting reagent. The concentration of a minor intermediate remains much smaller than the concentrations of any of the reactants. Detecting a major intermediate is usually viable, but the presence of minor intermediates may be difficult to prove.

Catalysts are usually defined as substances that accelerate a reaction without being consumed in it. Conversely, **inhibitors** slow down a process. Unfortunately, two IUPAC recommendations give very different definitions for the term catalyst:

- "A substance that participates in a particular chemical reaction and thereby increases its rate but without a net change in the amount of that substance in the system [8]."
- "A catalyst is a substance that increases the rate of a reaction without modifying the overall standard Gibbs energy change in the reaction [5];"

The first definition is very similar to the one used in this book. The second definition refers to the overall standard Gibbs energy change of the reaction, which is a very impractical because few experts consider determining Gibbs energies as a part of kinetic studies. It is also notable that the same set of recommendations [5] do not use Gibbs energies when the term inhibitor is defined.

More broadly, both catalysts and inhibitors are substances that appear in the rate equation but not in the overall stoichiometric equation: their presence influences the course of temporal changes, but their initial and final concentrations are the same. Yet, in catalysis research, a component is often called a catalyst even if it is consumed in the process. The rationale is that the amount of consumed "catalyst" is much lower than the amount of useful product formed. The consumption of the catalyst is usually called **inactivation** or catalyst **degradation** in this terminology. However, in many such instances, designating a substance as a catalyst reflects the objectives (or pre-conceptions) of the researcher rather than the observed reality.

It should be emphasized that catalysts and inhibitors actually take part in some reaction steps, as the principle of independent interactions makes it impossible to exert any effect without being involved in reactions. A catalyst is typically regenerated, and its reactions are often depicted in the form of a **catalytic cycle**. Drawing catalytic cycles are considered very instructive by many researchers, yet these pictures are often imprecise or even close to worthless from a kinetic point of view if they are used instead of properly determining rate equations.

An inhibitor must exert its effect in a more enigmatic way. Again, the principle of independent interactions requires that inhibitors cannot simply stop elementary reactions in which they do not participate. What they can do is to divert one of the reagents in a faster process. In this case, the product of the inhibited reaction will not form as at least one of the reactants is missing. A special kind of inhibitor is called **stabilizer**: this prevents the action of a catalyst that is an unwanted impurity.

The concentrations of products seldom appear in the rate equation. If they do, this fact can give rise **autocatalysis** or **autoinhibition**. In broader scientific terms, this phenomenon is often called feedback, which can be positive or negative. Autocatalysis can give rise to a number of highly unintuitive or exotic kinetic observations. The first experimental detection of an autocatalytic system was the reaction between permanganate ion and oxalic acid [6], which now serves as a classic example of positive feedback in chemistry.

1.4   Open Systems: Flow of Reactants and Photons

The rate equation is usually stated for a closed system. From a theoretical point of view, this is an important aspect as conservation equations depend on the lack of matter exchange with the surroundings. This condition is not true for **open systems**. Generally, any meaningful calculations in open systems require detailed knowledge about the matter flux between the reactor and its surroundings.

The effect of inflow and outflow is often most conveniently described as virtual reactions that have no reactant (for inflow) or product (for outflow). A special notation (Ø) is often used for these cases in stoichiometric equations. With these virtual reactions, the rate equation form given in Eq. (1.2) can often be used for open systems. Certain conservation laws may also apply, but their mathematical formulation may be much more difficult than in closed systems, because they are not direct consequences of physical conservation laws.

Open systems are seldom handled by books on chemical kinetics in a general fashion because the temporal behavior is typically governed by the particular form of openness and its external constraints. However, there is a highly popular type of open reactor called a **continuous stirred tank reactor** (very commonly abbreviated as **CSTR**), which is usually devoted special attention. This sort of reactor has an inlet, through which a particular mixture of reactants is led in and a spatially separate outlet, through which the reaction mixture is led away. The rate of inflow and outflow (in terms of volume/time) is identical and the volume of a CSTR is unchanged in time. The reactor is well stirred so that homogeneity is ensured. Therefore, a single concentration value for each species present is sufficient to characterize the state of the reactor. Flow means a direction-dependent external influence, so homogeneity in a CSTR must be actively maintained. To determine the time variation of concentrations in a CSTR, the rate equation must be supplemented by a flow term:

$$\frac{d[A_i]}{dt} = f_i([A_1], [A_2], \dots, [A_n]) + \frac{v_{\text{flow}}}{V}(c_i - [A_i]) \qquad (1.17)$$

In Eq. (1.17), v_{flow} is the volumetric rate of flow, which, as discussed in the previous paragraphs, is identical at the inlet and outlet, and V is the (time-independent) volume of the CSTR reactor. The ratio V/v_{flow} is the average residence time in the reactor. Its reciprocal $k_{\text{flow}} = v_{\text{flow}}/V$ is often interpreted as a first order flow rate constant. In fact, it is only this combination of parameters that is needed for the kinetic description, individual values of V and v_{flow} are not usually necessary. This quantity is also used in gas phase open reactors, where the name **space velocity**, or the less fortunate gas hourly space velocity (**GHSV**) is used.

Typically, chemical processes in a CSTR converge to a final, stationary state, which is unchanged in time and can be characterized without solving differential equations simply by finding the stationary concentrations ($[A_2]_\infty$) at which the rates of concentration change are zero for all species present:

$$0 = f_i([A_1]_\infty, [A_2]_\infty, \dots, [A_n]_\infty) + \frac{v_{\text{flow}}}{V}(c_i - [A_i]_\infty) \qquad (1.18)$$

However, the stationary concentrations may not be unambiguously defined by these equations, so several sets of stationary concentrations may exist. In addition, and especially if the chemical reactions show some autocatalysis, the solution of the CSTR rate equation in Eq. (1.17) may show oscillations rather than converging to a final $[A_2]_\infty$ value. Theoretical and experimental investigations in a CSTR still provide a fertile ground for finding exotic kinetic phenomena.

An extension of a CSTR is called a semi-open reactor, which only has an inlet. Some theoretical studies have been carried out in this type of reactor, the volume of which changes in time. This volume change is typically linear, which is equivalent to a time-independent rate of inflow, e.g., $V = V_0 + \beta t$. With the notations used in Eq. (1.17), the differential equation describing the semi-open reactor is as follows:

$$\frac{d[A_i]}{dt} + \frac{[A_i]}{V}\frac{dV}{dt} = f_i([A_1], [A_2], \ldots, [A_n]) + \frac{v_{\text{flow}}}{V}c_i \qquad (1.19)$$

Obviously, a semi-open reactor can only be operated for a limited time as there are certainly physical factors that limit the possible growth of volume.

An additional, very special type of a reactor is a photoreactor. It is a not particularly productive question of semantics whether this is a closed reactor or not. In any case, no flow terms are needed in a closed photoreactor, and conservation of matter can also be used in the usual way. The rate equation of a photochemically induced elementary step has a particular, non-mass action general form as a consequence of Beer's law. In this form, the rate is influenced by the intensity of light, and the ability of the entities to absorb light. It is easiest to state for a case when the illumination is monochromatic (i.e., only a single wavelength occurs):

$$f_i([A_1], \ldots, [A_n]) = f_{i,a}([A_1], \ldots, [A_n]) \times (1 - 10^{f_{i,b}([A_1], \ldots, [A_n])}) \qquad (1.20)$$

For polychromatic illumination (which means using several wavelengths simultaneously), individual wavelength contributions similar to that shown in Eq. (1.20) need to be integrated over the wavelength range. Photochemical systems might also involve non-photochemical reactions, these will contribute further power law terms to the individual f_i functions. Details and examples of the highly quantitative evaluation of photochemical reactions can be found in the literature [3, 4].

1.5 Experimental Design and Its Limitations

The essence of experiments in science is to carry out systematic tests to study natural phenomena in detail. This is no different in chemical kinetics. The design of such studies should ensure that the information content of the results is as high as possible. There are a few guidelines that must be followed in systematic kinetic studies. It is understood that physical reality (reactor size, solubility, mixing phenomena, sensitivity of analytical methods, etc.) often poses severe limitations on the possible experiments, yet the design should always go as far as these limits permit. In chemical kinetics, the primary goal is to learn the rate equation of a process through studying the rates of concentration change as a function of concentrations. The results and the conclusions drawn from them will always be limited to the range of concentrations studied, so the widest possible concentration ranges must be used in the studies to find the most general conclusions possible. As already remarked, there are very few experimental methods that give the rates of concentration change directly, so the rates are most often determined by monitoring the concentration as a function time. The rates of concentration change (and not only the concentrations) are also dependent on time, so this task is usually quite delicate. To avoid pitfalls, the researcher must be fully aware of the properties (and therefore the limitations) of the experimental method used for monitoring concentrations.

The **time resolution** of the monitoring method is of primary concern for kineticists. The method used for monitoring the process should have a **response time** that is considerably faster than the studied process itself. This is simply to ensure that the time dependence of the signal measured is characteristic of the investigated process and not the detection method used. Some compromise is not impossible here, but when the response time of the detection method is not significantly faster than the studied process, the resulting time-dependent experimental signal should be described as a **convolution** of concentration change and detector response characteristics.

Another important issue is the **concentration selectivity** of the monitoring method. Ideally, the concentrations of all species appearing in a chemical system should be followed selectively. In reality, that is seldom ever possible. Monitoring often gives an instrumental signal that is proportional to one of the concentrations, or a combination of concentrations. In general, efforts should be made to follow several different concentrations in processes. However, if time resolution and concentration selectivity are in conflict (i.e., methods that are excellent in one respect are poor from the other point of view), time resolution should be given priority. Kinetic data collected at insufficient time resolution are next to worthless, whereas insufficient concentration selectivity leaves a lot of room for mathematical techniques to extract useful (but usually limited) information. One of the reasons why kinetic evaluation based on (pseudo-)first order conditions is so robust (cf. Eq. (2.10) in Chap. 2) is that it can be used even when the concentration selectivity of the monitoring method is largely unknown. Another example is that the highly selective analytical technique gas chromatography coupled with mass spectrometry is basically not suitable for quantitative monitoring of kinetic measurements due to its poor time resolution, whereas notoriously unselective conductometry often yields useful data.

Detection methods may be offline or online. **Offline detection** has long traditions in chemical kinetics and is still often used in textbook examples. However, the value of offline methods as a source of quantitative information has diminished substantially. Offline detection means that the process is started somehow, allowed to run for a measured time, then some external influence is used to stop the process and the reaction mixture is submitted to analysis such as chromatography, classical titration, or some sort of spectroscopy (this is often referred to as **sampling**). The analysis often contains preparatory steps as well such as addition of reagents for derivatization, or physical separation. There are a multitude of potential problems with offline detection methods. First of all, a method should be found that stops the process reliably, but leaves the analyzed components unchanged. It must also be verified that preparatory steps do not interfere with components of interest—at least not in an unwanted or unknown way. The time resolution of offline methods is also obviously limited. On top of these problems, offline methods tend to be quite intensive in terms of human labor and consumption of chemicals. It is usually desirable to obtain at least 100 individual concentration points on each kinetic curve for reliable evaluation, doing this with offline methods typically involves prohibitively high costs. Therefore, offline methods are seldom used to obtain quantitative kinetic information in modern science. However, highly selective offline analysis may provide very useful qualitative kinetic information that is well worth the costs.

Online detection suits the needs of modern chemical kinetics much better. Such a method analyzes reaction mixtures without the need for sampling and easily records hundreds of individual points on a kinetic curve. The reaction is carried out within the sample compartment of the analysis instrument and monitored in real time. The measured physical property is often light absorption, light emission, some sort of electrode signal, conductivity, or magnetic properties. When online detection is used, it is important to make sure that the monitoring method does not interfere with the studied process. This might sound obvious, but is seldom given much thought in actual research. For example, during absorption measurements, the light used for analysis should not cause photochemical reactions. Failure to check such possible interferences led to erroneous conclusions in multiple cases [3]. Ion selective electrodes often have response times of seconds, which may not be fast enough compared to the studied process. The immersion of electrodes may contaminate the system with components that influence the studied process, which is usually unwanted, but is also often a source of such complications that makes evaluation impossible. It is very important that the experimenter should be aware of such possibilities and devise systematic tests to prove that unwanted influence does not corrupt the data collected during monitoring.

Another practical aspect of experiments in chemical kinetics is that the initiation of the reaction should be fast compared to the chemical process studied. Most often, initiation of the reaction is simple mixing, but sometimes it is provided by an external influence (a laser pulse or a short gamma ray pulse) that generates a reactant. Manual mixing typically takes a few seconds, whereas instrumental mixing in state-of-the-art stopped-flow instruments is usually complete in 1–2 ms. Reactions that feature characteristic time scales shorter than 1 ms cannot be studied if mixing is required. Laser pulses can be a lot shorter, but the fast initiation has to be verified in these cases as well. Again, the mathematical method of convolution can be used for borderline cases.

As in many other fields of science, the **principle of independent measurements** is an important one in chemical kinetics. This means that data should be obtained from several different sources, which are not connected to each other. A typical case in chemical kinetics is studying a reversible reaction. The equilibrium constant of such a process can be determined from both kinetic observations and a time-independent study of mixtures in which the equilibrium is already reached. In an ideal case, both of these studies should be carried out and the results compared to each other. Agreement between data from different sources gives a high level of confidence to conclusions, which is not limited to the actual value of an equilibrium constant, but is also valid for the entire model used.

As already pointed out, rates of concentration change are typically obtained from time-dependent concentrations. To calculate the rate, the derivative of the measured curve must be calculated. Numerical derivation, as a brief look to a textbook of numerical analysis will surely convince the reader, is highly difficult business and often plagued with uncertainties. The basic problem is that to learn the derivative, the concentration change should be measured over a short period of time, which means that the relative change is minor. Even with reasonably good precisions of concentration measurement, small differences in the value can only be measured with high relative error. Therefore, it is seldom practical to estimate a rate by dividing the concentration difference between two consecutive points on a kinetic curve by the time difference. It is much better to use a probe function to fit some portion of the experimental data, and use the analytically calculated derivative of the probe function as the reaction rate at a given time instant. More details will be presented in Sect. 2.3, which discusses fitting in general.

One important method to determine rate equations is based on measuring initial rates. This is also called van't Hoff differential method [11].[4] The essence is that the initial rates are determined in a number of different experiments and their dependence on the concentrations is studied. The original graphic evaluation is based on plotting the logarithm of initial rates as a function of the logarithm of the changed concentration, and the order of reaction is obtained as the slope of the straight line obtained in this plot. Today, nonlinear least squares fitting to untransformed data should be used, but it is still the functional form of the dependence of the initial rate on concentrations that must be explored.

The initial rate method will often yield a reasonably good idea about the rate law. Initial rates are those measured at $t = 0$. They have special significance, because all the concentrations are known at this time instant (initial concentrations), which is not the case at any later times as the studied process involves changes in the concentrations. Initial rate studies are typically designed in several different series of experiments with only one of the initial concentrations changed in each series and keeping the others constant. In power law type kinetics, a single series is suitable for finding the order of reaction with respect to the reagent whose concentration is changed. The products are not usually present in the initial mixture. This simplifies the exploration of the rate law somewhat. If the products are suspected to have some kinetic role in the process, they should be added intentionally to the initial mixture in a series of experiments to learn their effects. Unfortunately, adding intermediates to the initial mixture is seldom a viable method to study their effect.

Because of the mentioned uncertainty in the determination of rates, direct comparison of measured concentration data (or instrumental signals proportional to

---

[4]Dutch chemist Jacobus Henricus van't Hoff was awarded the first chemical Nobel prize in 1901. He is also generally considered to be the founding father of chemical kinetics.

them) with those predicted from the rate equation is more common than rate-based evaluations. In this technique, the experimenter has an initial guess for the form of the rate equation, which may come from chemical intuition, literature precedents or studying the initial rates. The theoretical kinetic curve is deduced by solving this rate equation, and then this function is fitted to the measured concentration data to learn whether it interprets them acceptably. An acceptable fit does not necessarily prove the correctness of the assumed rate equation, but an unacceptable fit clearly disproves it. This technique avoids the numerically quite uncertain calculation of rates from concentration data and also considers the entire time range in which data have been collected. Therefore, this is the major evaluation method used in today's kinetics. Solving a rate equation has a central role in this method. This is why Chap. 2 is dedicated solely to different solution methods.

In the next paragraphs, some specific information will be given for the use of light absorption to follow the kinetics of chemical reactions. Spectrophotometers are usually not particularly expensive instruments. Most chemicals interact with electromagnetic radiation in at least some part of the wavelength range between 200 and 800 nm, but the method is still uninvasive, so unlikely to interfere with the chemical processes under investigation. Therefore, light absorption is in principle suitable for monitoring most of the chemical reactions researchers might be interested in. In addition, absorbance measurements have a time resolution from heaven, picosecond time scales are not unheard of in measurements. As a result of the combination of these advantages, measurements in modern chemical kinetics are dominated by the spectrophotometric monitoring method. During these measurements, absorbance is used as a primary detected signal, which can be calculated from the degree of light absorption at a predetermined wavelength. Beer's law [1], stated here for the simultaneous presence of n different absorbing species, connects absorbances to concentrations in the following way:

$$A_\lambda = \sum_{i=1}^{n} \varepsilon_{i,\lambda} [A_i] \tag{1.21}$$

The parameter $\varepsilon_{i,\lambda}$ is called the **molar absorptivity** of component A_i at wavelength λ. Selectivity can be tuned by the selection of the wavelength, which can be utilized to collect a huge primary data set to characterize a process. Establishing the concentration selectivity sometimes can be a problem. The values of $\varepsilon_{i,\lambda}$ are best determined in independent measurements, but this is understandably very difficult to do for intermediates. Absorbance values basically reflect linear combinations of concentrations. This fact opened up new possibilities of data evaluation when the spread of personal computers made linear algebraic operations on large data matrices routinely available. For example, the number of different absorbing species contributing to an absorbance signal, i.e., the number of nonzero $\varepsilon_{i,\lambda}$ values in Eq. (1.21) can be determined by a numerical technique called matrix rank analysis (MRA). This book does not have the space or the conceptual need to present this method in detail, the interested reader is referred to the scientific literature [9, 10].

1.6   Systems Biology: Chemical Kinetics for Biologists

Different fields in science seem to have a hierarchy in both the complexity of phenomena studied and the depth of understanding achieved. This hierarchy is reflected by the saying that chemistry is best interpreted by reducing the explanations to basic physical phenomena, whilst biology is best interpreted by reducing the explanations to basic chemical phenomena.

During the last half century, biology has progressed steadily toward an understanding based on the molecular level. Fine chemical details of all substances important for life are now known, and in many examples, the molecular mechanisms of the physiological processes or diseases have been explored. Naturally, the initial attempts focused on structural aspects and experimental methods developed to solve similar problems in physics and chemistry have been successfully adapted to the needs of biology.

Biological systems are usually highly sensitive but can also respond to external effects with considerable efficiency. The potential for this sort of adaptability must of course be present in the structures of biologically important molecules, but the actual mechanism of such responses is typically based on the coordination of the functions of different elements. Today, it is clearly recognized that many important processes can only be interpreted on the level of systems: it is the coupling of different processes and mutual feedback that makes quick biological responses possible.

Systems biology has two central questions [7]:

• How do the molecular components interact in order to maintain their function?
• How do different cells interact each other to maintain higher levels of organization?

The interactions, both within and among cells, are communicated by certain chemicals. Therefore, an important element in answering these questions is to describe the temporal changes in the levels (or, to use the word of the chemists, concentrations) of the substances involved in the mechanisms through which the subsystems interact with each other. This problem is a very familiar one for chemical kineticists: the basic principles are identical, the equations are the same, and, the solution techniques must also be the same then.

The problems of interest is systems biology often show considerable complexity from a chemical point of view. Yet, it should not be forgotten that the time-honored methods developed in chemical kinetics can be used here as well because the fundamental nature of the problems is the same. The simulations of concentration changes in reactive systems do not depend on the purpose the concentrations are needed for. If simulations disagree with the observed reality, it must indicate limitations or errors in our knowledge in chemistry and biology alike. Although it is quite common to speak or write about the investigation of biochemical reaction networks, they are in no way different from chemical reaction networks.

A system can usually be characterized by its structure and its dynamics. The system structure is given by the listing the individual elements and also the possible interactions between them. The dynamics, on the other hand, is determined by the laws governing the specific interactions between these elements. In structural studies, time is not usually considered an important variable, whereas in dynamic studies, time is a central physical property. Even if there is a possible strong interaction between two elements of the system, this interaction will be without consequence if these elements never coexist.

The rise of systems biology understandably coincided with the rapid development of personal computers. Model calculations typically require considerable computational power, which was not routinely available before the 1990s. Another contributing factor to the rapid advancement of systems biology was that a critical mass of structural data had been collected about the biologically important molecules by the very beginning of the third millennium. So the structural knowledge was already at a sufficient level to build reasonably complex models, and the computational power was at hand too. By this time, chemical kinetics, the essence of which is understanding temporal changes in concentrations, was more than 100 years old as a science. Considerable experience had been gained about what useful information is present in the time course of concentrations and how this information can be deduced. This experience should be taken advantage of in systems biology as well.

References

1. Beer, A.: Bestimmung der Absorption des rothen Lichts in farbigen Flüssigkeiten. Ann. Phys. Chem. **86**, 78–88 (1852)
2. Érdi, P., Lente, G.: Stochastic Chemical Kinetics: Theory and (Mostly) Systems Biological Applications. Springer, New York (2014)
3. Fábián, I., Lente, G.: Light-induced multistep redox reactions: the diode-array spectrophotometer as a photoreactor. Pure Appl. Chem. **82**, 1957–1973 (2010)
4. Gombár, M., Józsa, É., Braun, M., Ősz, K.: Construction of a photochemical reactor combining a CCD spectrophotometer and a LED radiation source. Photochem. Photobiol. Sci. **11**, 1592–1595 (2012)
5. Laidler, K.J.: Glossary of terms used in chemical kinetics, including reaction dynamics. Pure Appl. Chem. **68**, 149–192 (1996)
6. Laidler, K.J.: The kinetics of the reaction between potassium permanganate and oxalic acid. J. Am. Chem. Soc. **54**, 2597–2597 (1932)
7. Lecca, P., Laurenzi, I., Jordan, F.: Determinsitic versus Sotchastic Modeling in Biochemistry and Systems Biology. Woodhead Publishing, Cambridge (2013)
8. Muller, P.: Glossary of terms used in physical organic chemistry. Pure Appl. Chem. **66**, 1077–1184 (1994)
9. Peintler, G., Nagypál, I., Jancsó, A., Epstein, I.R., Kustin, K.: Extracting experimental information from large matrixes. 1. A new algorithm for the application of matrix rank analysis. J. Phys. Chem. A **101**, 8013–8020 (1997)

10. Peintler, G., Nagypál, I., Epstein, I.R., Kustin, K.: Extracting experimental information from large matrices. 2. model-free resolution of absorbance matrices: M^3. J. Phys. Chem. A **106**, 3899–3904 (2002)
11. van't Hoff, M.J.H.: Etudes de dynamique chimique. Rec. Trav. Chim. Pays-Bas **3**, 333–336 (1884)

Chapter 2
Solving Rate Equations

The rate equation gives the rates of concentration change as a function of the concentrations themselves. A rate equation is said to be solved if a suitable function is found that satisfies the equation and gives the concentrations for each reaction time. It is trivial but not always clearly recognized that time can be the only independent variable in this solution: the time-dependent concentrations of other species cannot appear in it. The parameters of this function are usually the rate constants and initial concentrations. A common, but more confusing convention calls the rate equation **differential rate equation**, whilst the solution is termed **integrated rate equation**. Although the rate equation is certainly a differential equation and integration is typically needed to solve it, this wording does not do justice to the mathematical efforts needed to find the solution. In addition, this alternative convention also confuses equations with function definitions, which are usually separate concepts in mathematics and computation science.

The rate equation is always a **first order differential equation**, i.e., only first derivatives of the functions appear in it. It also has the important property of being **autonomous**: the independent variable (time) never appears directly in the rate equation, it exerts its effects exclusively through the concentrations. This is usually a favorable property from a mathematical point of view but also a very obvious one in a physical sense as it basically expresses the fact that the laws governing concentration changes (similarly to other laws of nature) do not depend on time. A further characteristic of the structure of the rate equation is that it gives the derivatives as an explicit function of the concentrations. Finally, a rate equation is typically nonlinear because the functions $f_i([A_1], [A_2], \ldots, [A_n])$ are nonlinear.

An **analytical solution** is a function that satisfies the rate equation and the initial conditions, which are given by the concentration values at $t = 0$. In contrast, a **numerical solution** is usually a huge collection of discreet time-concentration points that is estimated based on the rate equation. It is not uncommon to speak of **approximate analytical solutions** for cases when a function is found that is not the

© Gábor Lente 2015
G. Lente, *Deterministic Kinetics in Chemistry and Systems Biology*,
SpringerBriefs in Molecular Science, DOI 10.1007/978-3-319-15482-4_2

solution of the rate equation, but is close to it in some sort of measure. To avoid confusion, it would be best to drop the word "analytical" in this case and stick to the term approximate solution.

An analytical solution can typically be found only for relatively simple cases. However, when such an analytical solution is found, it clearly takes precedence over any numerical solution. This is why the following sections devote a lot of space to known analytical solutions.

2.1 Analytical Possibilities

2.1.1 Single-Concentration Rate Equations

The simplest possible class of rate equations only contains the concentration of a single species. In this case, function $f_1([A_1])$ also only has one independent variable. There is quite a good chance that a rate equation in this class can be solved analytically. At first sight, this class might seem utterly insignificant as no chemical process can be imagined that only contains a single species. Yet, a surprisingly high number of more complicated (and, of course, more realistic) rate equations can be transformed into a single-concentration variant using the law of mass conservation. As a consequence, single-concentration rate equations have a dominant role in classical chemical kinetics.

A general method to solve the differential equations in this class is called **separation of variables**. Variables in this phrase mean concentration (dependent) and time (independent). Separation is particularly easily achieved as time does not even appear in the equation explicitly: all that needs to be done is to divide both sides of the rate equation by $f_1([A_1])$. Subsequent integration of both sides gives rise to a form of the solution[1]:

$$\int_{[A_1]_0}^{[A_1]_t} \frac{1}{f_1([A_1])} d[A_1] = t \tag{2.1}$$

To calculate $[A_1]_t$ from this equation, one first needs to find the definite integral on the left-hand side, then rearrange the resulting equation so that it gives an explicit formula for the concentration (in mathematics, this is called function **inversion**).

---

[1]In a common description of this technique, the equation is said to be "multiplied" by the term dt and then "integrated." This method is often ridiculed by mathematicians, who point out that dt is part of a symbol on the left side, which does not have anything to do with division. Instead of joining the laughter, it is probably better to think a little bit about the fact that the method always gives the correct solution of the problem. In fact, this method also has rigorous mathematical background. It is called **nonstandard analysis** [16], where derivatives can be defined as a ratio of two infinitesimal quantities. But it is true that those who practice this method rarely have any ideas of the existence on nonstandard analysis.

If the first step (integration) cannot be completed, the rate equation cannot be solved analytically. However, failure to accomplish the second step means only that the solution found is implicit. In this case, as Eq. (2.1) is already explicit for t, **swapping the independent and dependent variables** is a useful trick. If concentration and time are swapped, the time passed to achieve a predetermined concentration can be calculated instead of determining the concentration at a given time.

The simplest member of the simplest class of rate equations is the **power law rate equation with a single concentration**:

$$\frac{d[A_1]}{dt} = -k[A_1]^\alpha \tag{2.2}$$

Because of traditions, some confusion may arise about the values of k for positive integer values of α. The source of this confusion is that such α values make it possible to give this rate equation in a kinetic mass action form:

$$\alpha A_1 \xrightarrow{k_1} \cdots \tag{2.3}$$

In this case, a consistent use of the conventions results in the appearance of α as a stoichiometric coefficient in the rate, so that $k = \alpha k_1$. Yet, this convention is often forgotten. Therefore, when reading the literature, it is imperative to check the definition of the rate constant, especially for $\alpha = 2$.

It must not be left without notice that this rate equation does not give a zero rate at $[A_1] = 0$ for non-positive orders of reaction (i.e., $\alpha \leq 0$). This should be kept in mind when using the general solution of Eq. (2.2), which is given as follows:

$$[A_1] = \left([A_1]_0^{1-\alpha} + (\alpha - 1)kt\right)^{\frac{1}{1-\alpha}} \quad \text{for} \quad \alpha \neq 1 \tag{2.4}$$

For $\alpha = 1$, the solution has a different form:

$$[A_1]_t = [A_1]_0 e^{-k_1 t} \quad \text{for} \quad \alpha = 1 \tag{2.5}$$

For $\alpha \geq 1$, the solutions tend to 0 asymptotically, so they decrease monotonously without ever reaching 0. For $\alpha < 1$, however, the concentrations reach the value of 0 at finite time, which gives rise to a **critical time**:

$$t_{\text{crit}} = \frac{[A_1]_0^{1-\alpha}}{(1-\alpha)k} \tag{2.6}$$

The source of this criticality is the fact the derivation of Eq. (2.1) involved a division by $f_1([A_1])$, which takes the value of 0 at the critical time. At this point, the function $[A_1]_t$ does not have a second derivative if $0 < \alpha < 1$ (break point occurs on the first derivative of the kinetic trace). For $\alpha = 0$, the curve does not have a first derivative (break point on the kinetic trace) at this point. For $\alpha < 0$, the critical time represents a point of **singularity** at which the derivative is undefined.

Among the possible values of α, 1 is the most common by far. The curve itself is called a **first order curve** or **exponential curve**, its equation is given separately in Eq. (2.5). A further case of high importance is the **second order trace** at $\alpha = 2$:

$$[A_1]_t = \frac{[A_1]_0}{1 + [A_1]_0 k t} \tag{2.7}$$

Furthermore, **zeroth order traces** with $\alpha = 0$ are sometimes found:

$$[A_1]_t = 0.5|[A_1]_0 - kt| + 0.5([A_1]_0 - kt) \tag{2.8}$$

The reader may be surprised to see **absolute value** signs (|) appear in this equation: it is some resourceful use of mathematics to ensure that the formula can be used for any value of t including those after t_{crit}.

Finally, **half-order traces** with $\alpha = 0.5$ might have occasional significance, and can be written using the trick with absolute values again:

$$[A_1]_t = \left(0.5|\sqrt{[A_1]_0} - 0.5kt| + 0.5(\sqrt{[A_1]_0} - 0.5kt)\right)^2 \tag{2.9}$$

The kinetic curves given in Eq. (2.4) have two parameters, k and $[A_1]_0$. Both of these are **scaling parameters**, which define the concentration and time scales, but do not influence the shapes of the kinetic curves. So, all exponential curves have the same shape, the only difference is the scaling. The same is true for any other single-concentration power law kinetic curve. Figure 2.1 shows the shapes of these kinetic

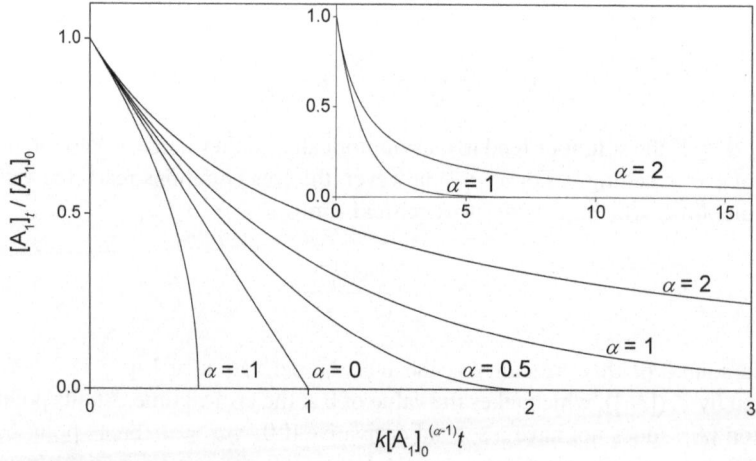

Fig. 2.1 Scaled kinetic traces of power-law rate equations in Eq. (2.2) with various values of the order of reaction α

curves for five different values of α. The characteristic shapes of these traces can be recognized by just looking at them. Scaling could be done in many different ways. Figure 2.1 uses what could be considered natural: the concentration unit is $[A_1]_0$, whereas the time unit is $k^{-1}([A_1]_0)^{1-\alpha}$. This scaling ensures that all the different curves shown have identical initial rates. This is a useful convention for preparing such comparative graphs, and can always be attained by setting the time unit to $[A_1]_0/f_1([A_1]_0)$ in single-concentration rate equations.

The natural time unit does not depend on the initial concentrations for $\alpha = 1$. Knowledge of the initial concentration is not necessary to determine the rate constant k in this case, the reciprocal of which is also called **lifetime**. Moreover, a **shift in time** (for example $t = t' + \Delta$) does not influence the determination of the rate constant, it merely gives a different value for the initial concentration ($= [A_1]_0 e^{-k\Delta}$). Therefore, the rate constant can be determined for an exponential curve for which neither the initial concentration nor the initial time is known precisely.

It is usually advisable to fit the measured signal directly rather than calculating concentrations first from the signal. If a signal Y is a linear combination of the concentrations, then the exponential curve for the observed signal is as follows:

$$Y_t = Xe^{-kt} + E \tag{2.10}$$

In this equation, X is termed the **amplitude**, k is the first order rate constant, whereas E is called the **endpoint**. The initial reading is simply $A + E$.

Furthermore, exponential curves are also easily handled for cases when the observed signal is integrated in time or space (or both). Integration in time (in interval τ) is often a key question when the response time of the monitoring method is not much faster than the process studied. The **integrated observation** is described as:

$$Y_t^\tau = \frac{1}{\tau} \int_t^{t+\tau} (Xe^{-ks} + E)ds = \frac{1 - e^{-k\tau}}{k\tau} Xe^{-kt} + E \tag{2.11}$$

Despite some apparent complexity, it must be recognized that Eq. (2.11) is still an exponential curve with rate constant k and endpoint E, it is only the amplitude that is influenced by the integration in time.

These facts make the kinetic methods based on exponential curve fitting highly **robust**. This robustness is the primary reason why kineticists, whenever possible, prefer finding conditions under which (pseudo-)first order kinetic curves are detected. Unfortunately, this strong preference is sometimes even driven to absurdity: a great many published works attempt to evaluate obviously non-exponential kinetic curves using first order fitting. Needless to say, no valid conclusions can be drawn from force-fitting first order rate constants to a non-exponential experimental trace.

Turning to a bit more complicated, but still single-concentration rate equations now, the following one, in which the rate is described by a rational function of the concentration, has high practical importance:

$$\frac{d[A_1]}{dt} = -\frac{k_a[A_1]}{k_b + [A_1]} \tag{2.12}$$

This is called the **Michaelis–Menten rate equation** and is often used in describing catalysis, especially in the field of enzyme kinetics [13]. The rate constants k_a and k_b are both positive (if $k_b = 0$, the equation would reduce to a zeroth order rate equation). The exact solution is surprisingly seldom used in the literature:

$$[A_1]_t = k_b W\left(\frac{[A_1]_0}{k_b}e^{([A_1]_0 - k_a t)/k_b}\right) \tag{2.13}$$

The solution uses the **Lambert W function** (denoted W), which is the inverse of the xe^x function. This is a three-parameter curve ($k_a, k_b, [A_1]_0$), so even when scaling is taken into account, curves may look different. A convenient scaling is $[A_1]_0$ as the concentration is unit and $(k_b + [A_1]_0)/k_a$ as the time unit. A third, dimensionless parameter combination $[A_1]_0/k_b$ characterizes the shapes of the curves. Figure 2.2 gives examples of the kinetic curves with different shape parameters.

These curves are often described to change their order of reaction from 0 in the beginning to 1 at the end. The rationale in this characterization is that at high values of concentration $[A_1]$, the rate does not depend on this concentration, whereas the

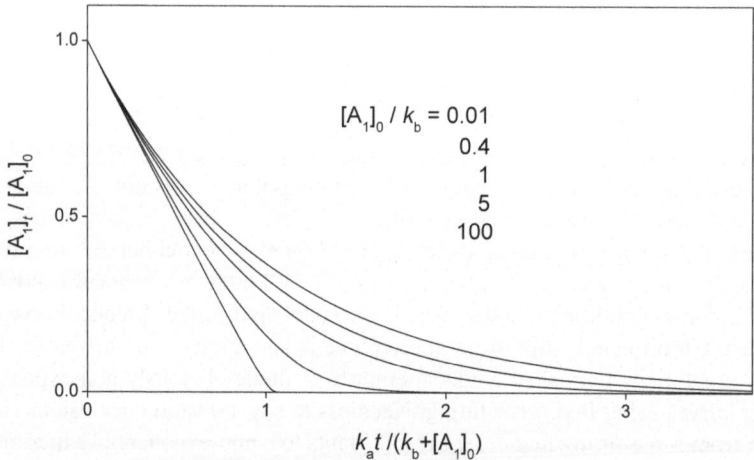

Fig. 2.2 Scaled kinetic traces based on Michaelis–Menten rate equation in (2.12) with different shape parameters, whose values are shown within the graph

rate is directly proportional to $[A_1]$ if its value is close to zero. The Lambert W function is not very commonly implemented in scientific softwares. In its absence, the trick of swapping time and concentration can still be used:

$$t = \frac{[A_1]_0 - [A_1]_t}{k_a} + \frac{k_b}{k_a} \ln \frac{[A_1]_0}{[A_1]_t} \tag{2.14}$$

In a similar, and sometimes significant rate equation, the order of reaction changes from 0.5 in the beginning to 1 at the end:

$$\frac{d[A_1]}{dt} = -\frac{k_a[A_1]}{k_b + \sqrt{[A_1]}} \tag{2.15}$$

The solution of this rate equation also uses the Lambert W function:

$$[A_1]_t = \left[k_b W \left(\frac{\sqrt{[A_1]_0}}{k_b} e^{(\sqrt{([A_1]_0} - k_a t/2)/k_b)} \right) \right]^2 \tag{2.16}$$

Further notable rate equations arise if terms of different power rate laws are summed. These are often called processes with parallel reaction paths, but their significance goes way beyond that, as will be shown later. A common and important case is when a first order and a second order term is summed:

$$\frac{d[A_1]}{dt} = -k_a[A_1] - k_b[A_1]^2 \tag{2.17}$$

As usual, separation of variables gives a straightforward solution:

$$[A_1]_t = \frac{k_a[A_1]_0 e^{-k_a t}}{k_a + k_b[A_1]_0 - k_b[A_1]_0 e^{-k_a t}} \tag{2.18}$$

This is a three-parameter curve, similarly to Eq. (2.16). The scaling parameters are best selected to be $[A_1]_0$ as the concentration unit and $|1/(k_a + k_b[A_1]_0)|$ as the time unit. The dimensionless shape parameter is $k_b[A_1]_0/k_a$. This formula reduces to a first order reaction by setting $k_b = 0$. However, $k_a = 0$ and $k_2[A_1]_0 = -k_a$ are not possible. Figure 2.3 gives 12 characteristic curve shapes.

The traces with positive shape parameters in Fig. 2.3 are straightforward, they are cases when both k_a and k_b are positive. Interestingly enough, one of the two rate constants can be negative without rendering the rate equation physically meaningless. Furthermore, the absolute value sign in the convenient time scale is not accidental, as $(k_a + k_b[A_1]_0)$ may also be negative. The condition $-1 < k_b[A_1]_0/k_a < 0$ and $(k_a + k_b[A_1]_0) > 0$ gives rise to curves that are often called autocatalytic: these curves are denoted by a + sign in the superscript after the parameter value in Fig. 2.3. For $-1 < k_b[A_1]_0/k_a < -0.5$, the traces have accelerating time intervals. If $(k_a + k_b[A_1]_0) < 0$, any negative value for the

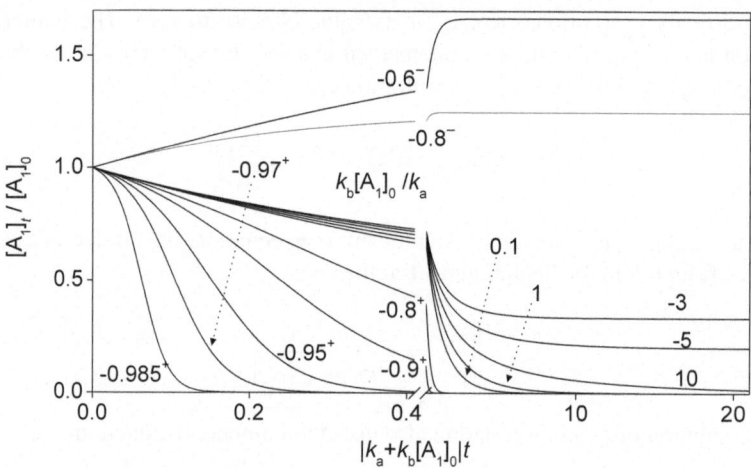

Fig. 2.3 Scaled kinetic traces based on rate Eq. (2.17) with different shape parameters, whose values are shown within the graph

shape parameter is meaningful. The solutions in this case do not tend to 0 final concentration values. The curves with $-1 < k_b[A_1]_0/k_a < 0$ describe an increase in concentration and are denoted by a $-$ sign in the superscript after the parameter value in Fig. 2.3.

A logical extension of the rate equation (2.17) is the addition of a third, zeroth order term:

$$\frac{d[A_1]}{dt} = -k_a[A_1] - k_b[A_1]^2 - k_c \qquad (2.19)$$

The solution breaks down to several different possibilities depending on the values of the parameters. If $4k_bk_c > k_a^2$, the solution is:

$$[A_1]_t = \frac{\sqrt{4k_bk_c - k_a^2}}{2k_b} \, \mathrm{tg}\left(-\frac{\sqrt{4k_bk_c - k_a^2}}{2}t + \mathrm{arctg}\,\frac{2k_b[A_1]_0 + k_a}{2}\right) - \frac{k_a}{2k_b} \qquad (2.20)$$

On the other hand, if $4k_bk_c < k_a^2$, the solution takes the following form:

$$[A_1]_t = \frac{k_a(Xe^{-t\sqrt{k_a^2 - 4k_bk_c}} - 1) + \sqrt{k_a^2 - 4k_bk_c}(Xe^{-t\sqrt{k_a^2 - 4k_bk_c}} + 1)}{2k_b(Xe^{-t\sqrt{k_a^2 - 4k_bk_c}} - 1)} \qquad (2.21)$$

In this equation, X is an auxiliary variable defined as:

$$X = \frac{k_b[A_1]_0 + k_a - \sqrt{k_a^2 - 4k_bk_c}}{k_b[A_1]_0 + k_a + \sqrt{k_a^2 - 4k_bk_c}} \tag{2.22}$$

Finally, if $4k_bk_c = k_a^2$ (or $k_c = k_a^2/(4k_b)$), the solution assumes a simpler form:

$$[A_1]_t = \frac{[A_1]_0 + k_a/(2k_b)}{k_b[A_1]_0t + k_at/2 + 1} - \frac{k_a}{2k_b} \tag{2.23}$$

These curves generally have four parameters, but the individual combinations that could serve as two scaling and two shape parameters are not obvious. If $k_a = 0$, the given solution is still useful, but the same is not true for $k_b = 0$. The solution of the case with $k_c = 0$ is already given in Eq. (2.18). The two-term rate equation that combines a zeroth and a first order term is handled separately:

$$\frac{d[A_1]}{dt} = -k_a[A_1] - k_c \tag{2.24}$$

The solution is stated in a relatively simple form:

$$[A_1]_t = \left([A_1]_0 + \frac{k_c}{k_a}\right)e^{-k_at} - \frac{k_c}{k_a} \tag{2.25}$$

A closer look at this formula will reveal to the reader that this is actually an exponential curve that is shifted along the concentration axis. Negative values of k_c also make sense in this case, especially for describing reversible reactions. There is also a critical time for positive values of k_c when the concentration assumes 0, $t_{crit} = (1/k_a)\ln(k_a[A_1]_0/k_c + 1)$.

Notably, some rate equations containing more than one concentration can be rewritten into single-concentration equations. A prime example is the so-called mixed second order equation:

$$A_1 + A_2 \xrightarrow{k_1} A_3(+\cdots) \tag{2.26}$$

The rate equation is:

$$\frac{d[A_1]}{dt} = \frac{d[A_2]}{dt} = -k_1[A_1][A_2] \tag{2.27}$$

Mass conservation ensures that $[A_2]_t = [A_1]_t - [A_1]_0 + [A_2]_0$ holds. Therefore, the rate equation can be rewritten into a single concentration form:

$$\frac{d[A_1]}{dt} = \frac{d[A_2]}{dt} = -k_1[A_1]^2 + k_1([A_2]_0 - [A_1]_0)[A_1] \tag{2.28}$$

This is the same rate equation as shown in Eq. (2.17) with $k_b = k_1$ and $k_a = k_1([A_1]_0 - [A_0]_0)$. Yet, as the mixed second order rate equation is very important, some further considerations will be presented about it. If it also involves a stoichiometric ratio different from 1:1, the process itself is usually represented as follows:

$$\nu_1 A_1 + \nu_2 A_2 \longrightarrow A_3(+\cdots) \qquad (2.29)$$

The rate equation corresponding to this process is then given as:

$$\frac{1}{\nu_1}\frac{d[A_1]}{dt} = \frac{1}{\nu_2}\frac{d[A_2]}{dt} = -k_1[A_1][A_2] \qquad (2.30)$$

Combining the general strategy outlined above and the solution already given in Eq. (2.18) yields the following final formula for the concentration of A_1:

$$[A_1]_t = \frac{(\nu_1[A_2]_0 - \nu_2[A_1]_0)[A_1]_0 e^{-k_1(\nu_1[A_2]_0 - \nu_2[A_1]_0)t}}{\nu_1[A_2]_0 - \nu_2[A_1]_0 e^{-k_1(\nu_1[A_2]_0 - \nu_2[A_1]_0)t}} \qquad (2.31)$$

A notable exception is the case $\nu_2[A_1]_0 = \nu_1[A_2]_0$. This is fully analogous to the $\alpha = 2$ case in Eq. (2.2), and the corresponding solution is very similar to the one given in Eq. (2.7) except the presence of the stoichiometric coefficient:

$$[A_1]_t = \frac{[A_1]_0}{1 + \nu_2[A_1]_0 k_1 t} \qquad (2.32)$$

The reader can test her or his own understanding of the kinetic background of this formula by thinking about the question why the stoichiometric coefficient ν_1 does not appear in it.

It is not very trivial to see, but Eq. (2.31) actually describes curves with three essential parameters. Two of these are scaling parameters: initial concentration $[A_1]_0$ serves as the concentration unit, whereas the reciprocal of the composite parameter $k_1(\nu_1[A_2]_0 - \nu_2[A_1]_0)$ is a time unit. The third parameter, which is best selected as the dimension-free ratio $\nu_1[A_2]_0/(\nu_2[A_1]_0)$ is a shape parameter. Figure 2.4 displays a few examples of curves with different shape parameters (Fig. 2.3 already shows some relevant curves, but the rate equation there is more general and the scaling is somewhat different). From a practical point of view, it is also sufficient to limit the considerations to $\nu_1[A_2]_0/(\nu_2[A_1]_0) \geq 1$ so that A_1 is the limiting reagent that is completely used up, and the concentration of A_2 can be calculated form mass balance. In the opposite case, swapping A_1 and A_2 would lead back to this mathematical description.

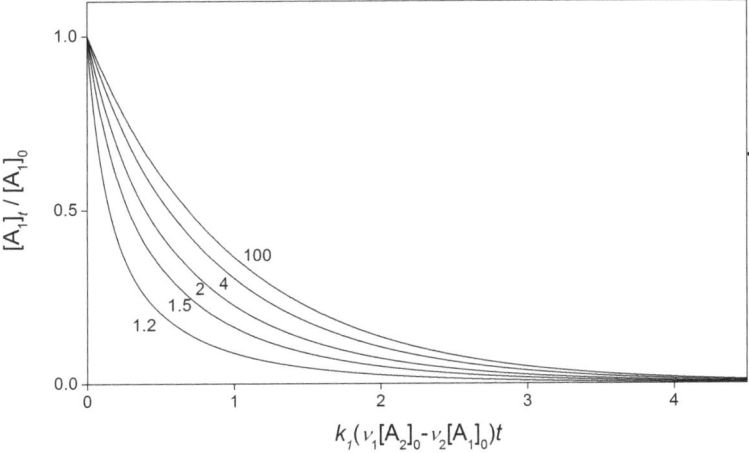

Fig. 2.4 Scaled kinetic traces based on Eq. (2.31) with different shape parameters

Another example in which a multiple-concentration rate equation can be simplified using mass balance is provided by the simplest autocatalytic rate equation including a direct term:

$$A_1 \longrightarrow A_2 \qquad (2.33)$$

The rate equation of this process is given as:

$$\frac{d[A_1]}{dt} = -\frac{d[A_2]}{dt} = -k_1[A_1] - k_2[A_1][A_2] \qquad (2.34)$$

Mass conservation ensures that $[A_2]_t = [A_1]_0 + [A_2]_0 - [A_1]_t$. Therefore, the rate equation can be rearranged into the following form:

$$\frac{d[A_1]}{dt} = -(k_1 + k_2[A_1]_0 + k_2[A_2]_0)[A_1] + k_2[A_1]^2 \qquad (2.35)$$

Again, this is the same as the rate equation shown in Eq. (2.17) with $k_b = -k_2$ and $k_a = k_1 + k_2[A_1]_0 + k_2[A_2]_0$. Some sample solutions are given in Fig. 2.3.

The following rate equation is the simplest photochemical rate equation, but receives very little attention in practice:

$$\frac{d[A_1]}{dt} = -k_a(1 - e^{-k_b[A_1]}) \qquad (2.36)$$

This describes a purely photochemical process for a case when the reaction is induced by monochromatic irradiation at which only the photoactive component absorbs light. These conditions are not uncommon in photochemistry. The term showing the concentration in the exponential function stems from Beer's law, which generally describes light absorption (see Eq. (1.21)). An analytical solution can be found for this case as well:

$$[A_1]_t = \frac{1}{k_b} \ln\left[1 + e^{-k_a k_b t} e^{k_b [A_1]_0} \left(1 - e^{-k_b [A_1]_0}\right)\right] \qquad (2.37)$$

This is again a three-parameter curve, where $k_a k_b$ and $[A_1]_0$ are scaling parameters, whereas $k_b [A_1]_0$ is a shape parameter. More complicated cases are also frequently encountered in photochemistry, but analytical solutions are typically very difficult, if not impossible to find because the concentration appears simultaneously in the exponential functions and their multiplication terms.

All the rate equations dealt with thus far were valid for irreversible reactions, although allowing negative values for the formal rate constant values also accommodates reversible cases, as will be shown later in this section. In reversible reactions, the initial substance A_1 can also be formed, and not only consumed.

Before dwelling on truly reversible chemical reactions, exchange reactions will be discussed at some length. An exchange process is commonly represented in the form of the following chemical scheme:

$$A_1^* + A_2 \rightleftharpoons A_1 + A_2^* \qquad (2.38)$$

The sign * indicates some sort of a label, which is usually assumed not to interfere with any of the reactions of the chemical species. In the case of isotopes, a modifying influence of the label is usually called isotope effect (see Sect. 4.4), and it is negligible except when hydrogen isotopes are substituted. Because of the negligible effect of the label, R_{ex}, the rate of exchange, can be interpreted as a single, time-independent quantity in such an experiment as the rate itself does not differentiate between A_1^* and A_1. It is usually useful to work with isotopic abundances x_1 and x_2 in molecules in this scheme, which can be defined as:

$$x_1 = \frac{[A_1^*]}{[A_1^*] + [A_1]} = \frac{[A_1^*]}{[A_1]_T} \qquad x_2 = \frac{[A_2^*]}{[A_2^*] + [A_2]} = \frac{[A_2^*]}{[A_2]_T} \qquad (2.39)$$

Concentrations $[A_1]_T$ and $[A_2]_T$ are the total concentration of species A_1 and A_2, which remain unchanged in time. The differential equation governing the time evolution of isotopic abundance x_1 is:

$$\frac{dx_1}{dt}[A_1]_T = -R_{ex}x_1(1 - x_2) + R_{ex}x_2(1 - x_1) = -R_{ex}x_1 + R_{ex}x_2 \qquad (2.40)$$

A similar equation for x_2 is:

$$\frac{dx_2}{dt}[A_2]_T = -R_{ex}x_2(1-x_1) + R_{ex}x_1(1-x_2) = R_{ex}x_1 - R_{ex}x_2 \quad (2.41)$$

Surprising as it may be, this is not a true rate equation as the exchange rate R_{ex} appears in it without giving the concentration dependence. The rationale in this formulation is that the quantity R_{ex} does not change during the course of a single kinetic experiment.

Adding Eqs. (2.40) and (2.41) shows that $x_1[A_1]_T + x_2[A_2]_T$ is independent of time, which is actually mass conservation for the label. This enables calculation of x_2 from the value of x_1:

$$x_2 = \frac{[A_1]_T}{[A_2]_T}(x_{1,0} - x_1) + x_{2,0} \quad (2.42)$$

The equation can thus be restated with a single dependent variable, and then the equation is solved to yield:

$$x_{1,t} = (x_{1,0} - x_\infty)\exp\left(-\frac{[A_1]_T + [A_2]_T}{[A_1]_T[A_2]_T}R_{ex}t\right) + x_\infty \quad (2.43)$$

The final value of the isotopic abundance is common to the two species:

$$x_\infty = \frac{x_{1,0}[A_1]_T + x_{2,0}[A_2]_T}{[A_2]_T + [A_1]_T} \quad (2.44)$$

A fully symmetric equation gives the solution for $x_{2,t}$. Equation (2.43) means that the time dependence of isotopic abundance is always described by an exponential curve, no matter what the rate equation of the exchange process is. The rate equation itself can be established by determining the first order rate constants of the exponential curves and studying their dependence on the concentration of the species. The approach presented here and specifically Eq. (2.43) are typically referred to as the McKay equation [11, 12].

This exchange example leads further to the discussion of single-step **reversible reactions**. The general strategy is to use the conservation of mass to give all concentrations as a function of a single selected concentration. The simplest example is a reaction that is first order in both directions:

$$A_1 \underset{k_2}{\overset{k_1}{\rightleftharpoons}} A_2 \quad (2.45)$$

The rate equation of such a process is:

$$\frac{d[A_1]}{dt} = -\frac{d[A_2]}{dt} = -k_1[A_1] + k_2[A_2] \quad (2.46)$$

Despite the fact that two concentrations appear in it, this is a single-concentration rate equation as $[A_1]$ and $[A_2]$ are connected through mass conservation: $[A_2]_t = [A_1]_0 + [A_2]_0 - [A_1]_t$. With this conservation equation, it is a matter of very simple algebra to derive the following form:

$$\frac{d[A_1]}{dt} = -(k_1 + k_2)[A_1] + k_2([A_1]_0 + [A_2]_0) \tag{2.47}$$

This rate equation is identical to the one given in Eq. (2.24) with $k_a = k_1 + k_2$ and $k_b = k_2([A_1]_0 + [A_2]_0)$. The solution is therefore an exponential function with a nonzero final value.

Another reversible process involves a first and a second order reaction:

$$A_1 \underset{k_2}{\overset{k_1}{\rightleftharpoons}} A_2 + A_3 \tag{2.48}$$

The rate equation describing this scheme is:

$$\frac{d[A_1]}{dt} = -\frac{d[A_2]}{dt} = -\frac{d[A_3]}{dt} = -k_1[A_1] + k_2[A_2][A_3] \tag{2.49}$$

The analytical solution of this equation is the same as given after Eq. (2.19) with $k_a = k_1 + k_2(2[A_1]_0 + [A_2]_0 + [A_3]_0)$, $k_b = -k_2$, and $k_c = -k_2([A_1]_0 + [A_2]_0)([A_1]_0 + [A_3]_0)$.

Another significant reversible process is when both the forward and reverse reactions are second order overall:

$$A_1 + A_2 \underset{k_2}{\overset{k_1}{\rightleftharpoons}} A_3 + A_4 \tag{2.50}$$

The rate equation is given as follows:

$$\frac{d[A_1]}{dt} = \frac{d[A_2]}{dt} = -\frac{d[A_3]}{dt} = -\frac{d[A_4]}{dt} = -k_1[A_1][A_2] + k_2[A_3][A_4] \tag{2.51}$$

Similarly to the previous reversible scheme, the analytical solution given after Eq. (2.19) can be used here with $k_a = k_1([A_2]_0 - [A_1]_0) + k_2(2[A_1]_0 + [A_3]_0 + [A_4]_0)$, $k_b = k_1 - k_2$, and $k_c = -k_2([A_1]_0 + [A_3]_0)([A_1]_0 + [A_4]_0)$.

One more trick should be introduced for reversible reactions at this point using the process shown in Eq. (2.50) as an example. Reversible processes lead to an equilibrium, in which all of the components reach a nonzero final concentration. It leads to considerably simplified algebra if the differential equation is stated using a new variable, which is the **distance from equilibrium**. In the example of Eq. (2.50), these equilibrium or final concentrations are designated $[A_1]_\infty$, $[A_2]_\infty$, $[A_3]_\infty$, and $[A_4]_\infty$. For these, $k_1[A_1]_\infty[A_2]_\infty = k_2[A_3]_\infty[A_4]_\infty$ holds. The distance from the equilibrium is introduced as time dependent function x, and the individual concentrations can be given at any time with the following equation:

$$[A_1]_t = [A_1]_\infty + x_t \quad [A_2]_t = [A_2]_\infty + x_t$$
$$[A_3]_t = [A_3]_\infty - x_t \quad [A_4]_t = [A_4]_\infty - x_t \tag{2.52}$$

Substituting these functions into Eq. (2.51) and rearrangement yield the following differential equation for x:

$$\frac{dx}{dt} = -(k_1[A_1]_\infty + k_1[A_2]_\infty + k_2[A_3]_\infty + k_2[A_4]_\infty)x + (k_2 - k_1)x^2 \tag{2.53}$$

Therefore, the introduction of the distance form equilibrium (x) offers the advantage that the differential equation describing it, which is identical to Eq. (2.17) with $k_a = k_1[A_1]_\infty + k_1[A_2]_\infty + k_2[A_3]_\infty + k_2[A_4]_\infty$ and $k_b = k_2 - k_1$, is simpler. The price of this (relative) simplicity is that the equilibrium concentrations have to be obtained in independent equilibrium calculations. On a conceptual level, this method fully separates the equilibrium information and the kinetic information available from the concentration data.

An archetype of two parallel processes is the first order formation of two different products from the same initial substance. The scheme can be given as follows:

$$A_1 \xrightarrow{k_1} A_2$$
$$\tag{2.54}$$
$$A_1 \xrightarrow{k_2} A_3$$

It may be surprising, but this scheme also belongs to the group of single-concentration rate equations as A_2 and A_3 are only involved in the reaction as products:

$$\frac{d[A_1]}{dt} = -(k_1 + k_2)[A_1]$$
$$\frac{d[A_2]}{dt} = k_1[A_1] \tag{2.55}$$
$$\frac{d[A_3]}{dt} = k_2[A_1]$$

The solution is quite straightforward:

$$[A_1]_t = [A_1]_0 e^{-(k_1+k_2)t}$$
$$[A_2]_t = \frac{k_1}{k_1 + k_2}[A_1]_0 e^{-(k_1+k_2)t} \tag{2.56}$$
$$[A_3]_t = \frac{k_2}{k_1 + k_2}[A_1]_0 e^{-(k_1+k_2)t}$$

2.1.2   Multiple-Concentration Rate Equations

Multiple-concentration rate equations are those which cannot be simplified to a differential equation using only a single concentration, so they remain systems of simultaneous, coupled equations.

The end of the previous subsection showed that the simplest case of parallel processes, although involves two reactions with two rate constants, is in fact a single concentration rate equation as the reagent is common to the two irreversible processes. Therefore, the simplest multiple-concentration rate equations arise from consecutive processes.

A consecutive zeroth order process would be characterized by kinetic curves that are combinations of straight lines. Zeroth order kinetics is rare, so a case of two consecutive reactions both with zeroth order kinetics would be an extreme rarity, although still not without an experimental example. In some cases, the multistep oxidation reactions of the dithionate ion produce this unique phenomenon [8].

The class of first order reaction networks is among the few multiple-concentration rate equations for which the analytical solution can be given. The simplest such system is composed of two consecutive irreversible first order processes:

$$A_1 \xrightarrow{k_1} A_2 \xrightarrow{k_2} A_3 \tag{2.57}$$

The rate equation is:

$$\frac{d[A_1]}{dt} = -k_1[A_1]$$

$$\frac{d[A_2]}{dt} = k_1[A_1] - k_2[A_2] \tag{2.58}$$

$$\frac{d[A_3]}{dt} = k_2[A_2]$$

This scheme has two concentrations that should be handled in differential equations as $[A_3]$ can always be calculated from mass balance. The solution can be stated in terms of combinations of exponential functions. The formula for A_1 is a single exponential function, it is the solution of a single-concentration rate equation. The concentrations of A_2 and A_3, on the other hand, are described by functions called **biexponential** or **double exponential** functions:

$$[A_1]_t = [A_1]_0 e^{-k_1 t}$$

$$[A_2]_t = \frac{[A_1]_0 k_1}{k_1 - k_2}(e^{-k_2 t} - e^{-k_1 t}) + [A_2]_0 e^{-k_2 t} \tag{2.59}$$

$$[A_3]_t = [A_1]_0 + [A_2]_0 + [A_3]_0 + \frac{[A_1]_0 k_2}{k_1 - k_2} e^{-k_1 t} - \left(\frac{[A_1]_0 k_1}{k_1 - k_2} + [A_2]_0\right) e^{-k_2 t}$$

The formulas given for A_2 and A_3 cannot be used if $k_a = k_b (= k)$. In this case, the solutions take the following forms:

$$[A_2]_t = ([A_1]_0 kt + [A_2]_0)e^{-kt}$$

$$[A_3]_t = [A_1]_0 + [A_2]_0 + [A_3]_0 - ([A_1]_0 kt + [A_1]_0 + [A_2]_0)e^{-kt} \quad (2.60)$$

The concentration profile of A_2 shows a maximum in certain cases. The time at which this maximum occurs is given as:

$$t_{max} = \frac{1}{k_a - k_b} \ln \frac{\frac{k_a [A_1]_0}{k_a - k_b} + [A_2]_0}{\frac{k_b [A_1]_0}{k_a - k_b}} \quad (2.61)$$

If this formula gives a negative value for t_{max}, the concentration profile of A_2 decreases monotonously. In real experiments, typically $[A_2]_0 = [A_3]_0 = 0$ holds, which simplifies the solution somewhat. When the initial concentration of A_2 is zero, the concentration of this species is guaranteed to have a maximum and Eq. (2.61) is greatly simplified to assume the following form:

$$t_{max} = \frac{\ln(k_a / k_b)}{k_a - k_b} \quad (2.62)$$

In this example, species A_2 is the archetype of an intermediate, and the whole reaction is understood as a transformation of A_1 into A_3 through intermediate A_2. This scheme is used very often, so it will be dealt with in some more detail. The kinetic curves have two scaling parameters: an advantageous time unit is k_b^{-1}, whereas $[A_1]_0$ is a suitable concentration unit. It is usually a good choice to select the dimensionless combination k_a / k_b as a shape parameter. Figure 2.1 shows how the kinetic traces for A_2 and A_3 depend on this shape parameter. A_1 is not displayed as its dependence is single exponential. In fact, it does not have a shape parameter and unlike for the other two species, selecting k_a^{-1} as the time unit is more advantageous.

Curves in Fig. 2.5 show that the maximum concentration of the intermediate decreases as k_a / k_b increases. This is understandable as the rate constant of the process consuming the intermediate becomes larger with the increase of this ratio.

Another phenomenon to be noted is that the concentration profile of species A_3 features a region where the rate of formation increases. This is often referred to as an **induction time** (or incubation time or even lag phase). Autocatalytic curves for the formation of a product often have similar features, but the induction period is also observed there in the concentration of initial substance A_1. Strictly speaking, the initial rate of the formation of A_3 is zero as A_2 is not present at $t = 0$.

The double exponential function is very commonly used in experimental kinetics. Its form for an instrumental reading is:

$$Y_t = X_1 e^{-k_1 t} + X_2 e^{-k_2 t} + E \quad (2.63)$$

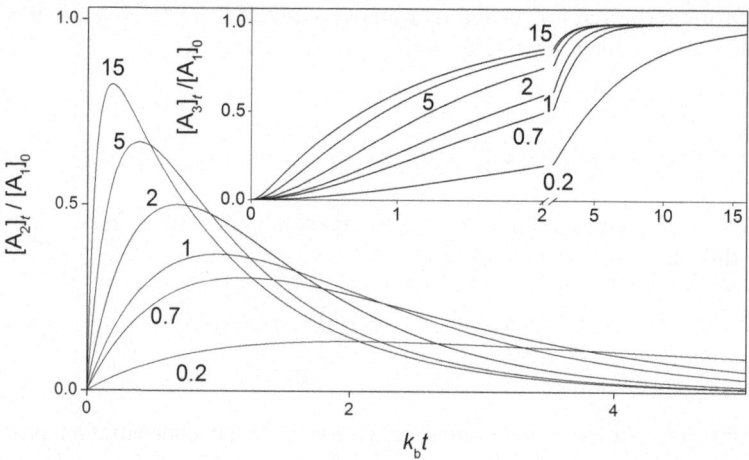

Fig. 2.5 Scaled kinetic traces based on Eq. (2.59) with different k_a/k_b shape parameters, the values of which are shown in the figure

In this equation, X_1 and X_2 are referred to as the first and second amplitudes (they can be positive or negative), k_1 and k_2 are the first and second rate constants, usually the higher given first (but that does not imply that the faster is always the first in the scheme!), whereas E is the endpoint. The initial reading is simply $X_1 + X_2 + E$.

A series of n irreversible first order reactions is represented by the following scheme:

$$\text{A}_1 \xrightarrow{k_1} \text{A}_2 \xrightarrow{k_2} \cdots \longrightarrow \text{A}_n \xrightarrow{k_n} \text{A}_{n+1} \tag{2.64}$$

The concentration of A_i $(i \leq n)$ in this scheme can usually be given as a combination of i exponential functions:

$$[\text{A}_i]_t = \sum_{j=1}^{i} C_{i,j} e^{-k_j t} \tag{2.65}$$

The coefficients $C_{i,j}$ can be calculated recursively in the following manner for any $i \leq n$:

$$C_{1,1} = [\text{A}_i]_0$$
$$C_{i,j} = \frac{k_{i-1}}{k_j - k_i} C_{i-1,j} \quad i > j \tag{2.66}$$
$$C_{i,i} = [\text{A}_i]_0 - \sum_{j=1}^{i-1} C_{i,j}$$

For the final product, species A_{n+1}, mass conservation gives a convenient way to calculate the concentration once the concentrations of the previous components are known:

$$[A_{n+1}]_t = [A_{n+1}]_0 + \sum_{j=0}^{n}([A_j]_0 - [A_j]_t) \qquad (2.67)$$

As in the case of single concentration rate equations, allowing the processes to be reversible will not change the double exponential nature of the solution, but will make the calculations more laborious. The most general case is when two consecutive reversible first order reactions follow each other, this is represented by the following scheme:

$$A_1 \underset{k_2}{\overset{k_1}{\rightleftharpoons}} A_2 \underset{k_4}{\overset{k_3}{\rightleftharpoons}} A_3 \qquad (2.68)$$

The solution is given by double exponential functions:

$$[A_i]_t = [A_i]_\infty + C_{i,1}e^{-\lambda_1 t} + C_{i,2}e^{-\lambda_2 t} \qquad (2.69)$$

In this equation, λ_1 and λ_2 are the nonzero eigenvalues of a 3×3 matrix composed of the rate constants k_1, k_2, k_3, k_4 (see later paragraphs about compartmental processes) and are given as follows:

$$\lambda_1 = -\frac{k_1+k_2+k_3+k_4}{2} + \sqrt{\frac{(k_1+k_2+k_3+k_4)^2}{4} - k_1 k_3 - k_1 k_4 - k_2 k_4}$$

$$\lambda_2 = -\frac{k_1+k_2+k_3+k_4}{2} - \sqrt{\frac{(k_1+k_2+k_3+k_4)^2}{4} - k_1 k_3 - k_1 k_4 - k_2 k_4} \qquad (2.70)$$

The equilibrium values of the concentrations can be given independently of the solution of the differential equation:

$$[A_1]_\infty = \frac{([A_1]_0 + [A_2]_0 + [A_3]_0)k_2 k_4}{k_1 k_4 + k_2 k_4 + k_1 k_3} \qquad [A_2]_\infty = \frac{([A_1]_0 + [A_2]_0 + [A_3]_0)k_1 k_4}{k_1 k_4 + k_2 k_4 + k_1 k_3}$$

$$[A_3]_\infty = \frac{([A_1]_0 + [A_2]_0 + [A_3]_0)k_1 k_3}{k_1 k_4 + k_2 k_4 + k_1 k_3} \qquad (2.71)$$

The constants appearing in Eq. (2.68) (which are the amplitudes of the double exponential functions) are given as follows:

$$C_{1,1} = \frac{\lambda_2 + k_1}{\lambda_2 - \lambda_1}([A_1]_0 - [A_1]_\infty) + \frac{k_2}{\lambda_2 - \lambda_1}([A_2]_\infty - [A_1]_0)$$

$$C_{1,2} = \frac{\lambda_1 + k_1}{\lambda_1 - \lambda_2}([A_1]_0 - [A_1]_\infty) + \frac{k_2}{\lambda_1 - \lambda_2}([A_2]_\infty - [A_1]_0)$$

$$C_{3,1} = \frac{\lambda_2 + k_4}{\lambda_2 - \lambda_1}([A_3]_0 - [A_3]_\infty) + \frac{k_3}{\lambda_2 - \lambda_1}([A_2]_\infty - [A_1]_0) \qquad (2.72)$$

$$C_{3,2} = \frac{\lambda_1 + k_4}{\lambda_1 - \lambda_2}([A_3]_0 - [A_3]_\infty) + \frac{k_3}{\lambda_1 - \lambda_2}([A_2]_\infty - [A_1]_0)$$

$$C_{2,1} = -C_{1,1} - C_{3,1} \qquad C_{2,2} = -C_{1,2} - C_{3,2}$$

Compartmental processes [4, 9] are usually general networks of first order reactions in a closed system, which involve n different chemical species $(A_1, A_2, \ldots, A_n)$, every one of which can convert to any other, i.e., chemical reactions are possible for all pairs of species present:

$$A_i \xrightarrow{k_{i,j}} A_j \qquad (2.73)$$

This sort of reaction is sometimes termed **conversion** to distinguish it from other types of possible processes in compartmental systems. For easy symbolic handling, the rate constant of the "self-conversion" of species A_i is also defined but is set to zero ($k_{i,i} = 0$).

From a chemist's point of view, a reactor can be made open with inflow and outflow by including processes that produce or consume A_i molecules without consuming or producing any other molecules. The outflow process, sometimes termed **degradation**, is still within the general framework of first order reaction networks. The same is not true for the inflow process, which would be represented by a constant term that does not depend on any of the concentrations (see later). The degradation process is typically represented by the following notation:

$$A_i \xrightarrow{k_i^{out}} \emptyset \qquad (2.74)$$

The notation $\emptyset$ is used to mean the absence of species on the product side, or, more precisely, the fact that molecule A_i leaves the reactor. There is also a whole family of steps, termed **catalytic production from a source**, in which species A_j is produced with a rate that is proportional to number of A_i molecules:

$$\emptyset \xrightarrow{k_{i,j}^{cat},A_i} A_j \qquad (2.75)$$

The notable case of $i = j$ here could be considered **self-reproduction** or autocatalytic formation. It should be noted that **catalytic degradation**, which would be a sort of equivalent of catalytic production, is not possible because a species cannot be degraded in a manner that is independent of its own presence.

Although the outflow process is often included in schemes as written in Eq. (2.74), this is not a conceptual necessity. In fact, in any system involving outflow, defining a $(n + 1)$th reservoir species, and including it as a common product of all outflow processes, will lead to a mathematically equivalent scheme.

In the absence of production steps, such an equivalent system will be closed. Open chemical reactors are sometimes conveniently formulated without the inflow or outflow processes described here in order to keep the number of molecules in the reactor finite. In this case, the **flow reactions** replace the contents of the reactor by the contents of the feed.

Another question arises about stoichiometry. All equations here are written with a set 1:1 stoichiometry for each process (meaning that for each molecule of reactant produced, there is one molecule of product formed) in the compartmental network. This is not necessarily the case in all examples and including stoichiometric coefficients may be needed. For 1:1 stoichiometry in a closed system without inflow and outflow, conservation of matter ensures that the sum of concentrations is always the same. With different stoichiometries but still in closed systems, a similar conservation relationship can be defined by using a suitable linear combination of concentrations.

A compartmental system is best characterized by matrix algebra and the rate constants are conveniently arranged in a form of a matrix, which is denoted $\underline{k}$ here:

$$\underline{k} = \begin{pmatrix} -k_1^{out} - \sum_{i=1}^{n} k_{1,i} & k_{2,1} + k_{2,1}^{cat} & \cdots & k_{n,1} + k_{n,1}^{cat} \\ k_{1,2} + k_{1,2}^{cat} & -k_2^{out} - \sum_{i=1}^{n} k_{2,i} & \cdots & k_{n,2} + k_{n,2}^{cat} \\ \vdots & \vdots & \ddots & \vdots \\ k_{1,n} + k_{1,n}^{cat} & k_{2,n} + k_{2,n}^{cat} & \cdots & -k_n^{out} - \sum_{i=1}^{n} k_{n,i} \end{pmatrix} \quad (2.76)$$

If the concentrations $[A_1], [A_2], \ldots, [A_n]$ are arranged into a vector $(\underline{c})$, the rate equation of a first order reaction network is simply given as follows:

$$\frac{d\underline{c}}{dt} = \underline{k}\underline{c} \quad (2.77)$$

The solution of this equation is most conveniently given in a matrix form:

$$\underline{c}_t = \text{expm}(\underline{k}t)\underline{c}_0 \quad (2.78)$$

In this formula, expm stands for the matrix exponential function, which is defined in a fashion that is fully analogous to the definition of the exponential function of real numbers. For any matrix $\underline{M}$, its exponential is given as follows:

$$\text{expm}(\underline{M}) = \sum_{i=0}^{\infty} \frac{1}{i!} \underline{M}^i \quad (2.79)$$

In Eq. (2.78), $\underline{c}_0$ represents the initial conditions, i.e., the values of concentrations at $t = 0$. The individual $[A_i]$ functions can be given using the eigenvalues of matrix $\underline{k}$. Let m be the number of different eigenvalues of matrix $\underline{k}$, $\lambda_1, \lambda_2, \ldots, \lambda_m$ the eigenvalues themselves, and $l_1, l_2, \ldots, l_m$ the multiplicities of these eigenvalues, in order. The following equation holds for the multiplicities:

$$n = \sum_{i=1}^{m} l_i \tag{2.80}$$

If the system is closed, the sums of all columns in matrix $\underline{\underline{k}}$ are zero, and the matrix itself is singular: at least one of its eigenvalues is zero. The solution can also be given without using the matrix exponential function, based on a combination of exponential functions and polynomials:

$$[A_i] = \sum_{j=1}^{m} \sum_{h=1}^{l_j} C_{i,j,h} t^{h-1} e^{\lambda_j t} \tag{2.81}$$

Complex numbers may arise as eigenvalues, but as all the elements of matrix $\underline{\underline{k}}$ are real, they can only appear in conjugate pairs. In this case, it is always possible to reformulate the solution using the real sine and cosine functions only, thus eliminating the need for using the complex exponential function. The values of constants $C_{i,j,h}$ can be given based on the initial conditions.

In the reactions representing the inflow, molecules are produced. These processes are sometimes termed **production from a source**:

$$\emptyset \xrightarrow{k_i^{in}} A_i \tag{2.82}$$

This will lead to a first order, inhomogeneous linear system of differential equations. The solution of this is relatively easily stated in a matrix form if the k_i^{in} values are arranged into a vector $\underline{k}^{in}$ as:

$$\underline{c}_t = \text{expm}(\underline{\underline{k}}t)(\underline{c}_0 - \underline{\underline{k}}^{-1}\underline{k}^{in}) + \underline{\underline{k}}^{-1}\underline{k}^{in} \tag{2.83}$$

Consecutive processes containing higher order reaction steps seldom have known analytical solutions. When the number of consecutive steps is only two, suitable analytical formulas can sometimes be found, but they often contain mathematical functions that are not generally preferred by chemists. Nevertheless, some of these analytical solutions will be given here.

A second order reaction followed by a first order process is described by the following scheme [10]:

$$2A_1 \xrightarrow{k_1} A_2 \xrightarrow{k_2} A_3 \tag{2.84}$$

The concentration of A_1 is obtained easily from a single-concentration rate law. The concentration of the intermediate is a more delicate matter:

$$[A_2]_t = \frac{4k_1[A_2]_0 + 2k_1[A_1]_0 - k_2 e^{-k_2/(2k_1[A_1]_0)} \mathrm{Ei}\left(\frac{k_2}{2k_1[A_1]_0}\right)}{4k_1} e^{-k_2 t}$$

$$+ \frac{-2k_1[A_1]_0 - k_2 e^{-k_2 t - k_2/(2k_1[A_1]_0)}(1 + 2k_1[A_1]_0)\mathrm{Ei}\left(k_2 t + \frac{k_2}{2k_1[A_1]_0}\right)}{4k_1(1 + 2k_1[A_1]_0)} \tag{2.85}$$

In this formula, $\mathrm{Ei}(x)$ is the exponential integral function, which is defined by the following integral:

$$\mathrm{Ei}(x) = -\int_{-x}^{\infty} \frac{e^{-z}}{z} dz \tag{2.86}$$

The scheme involving a mixed second order reaction and a subsequent first order process can also be solved analytically [10]:

$$A_1 + A_2 \xrightarrow{k_1} A_3 \xrightarrow{k_2} A_4 \tag{2.87}$$

The concentration of A_1 and A_2 is obtained readily from a single-concentration rate law and is given in Eq. (2.31). The general solution for intermediate A_3 is given as:

$$[A_3]_t = [A_3]_0 e^{-k_2 t} + [A_1]_0 e^{-k_2 t} - \frac{k_2[A_1]_0([A_2]_0 - [A_1]_0)H(0)}{[A_2]_0(k_1[A_1]_0 - k_1[A_2]_0 + k_2)} e^{-k_2 t}$$

$$+ \frac{[A_1]_0([A_2]_0 - [A_1]_0)e^{([A_1]_0 - [A_2]_0)k_1 t}}{([A_1]_0 e^{([A_1]_0 - [A_2]_0)k_1 t} - [A_2]_0)} + \frac{k_2[A_1]_0([A_2]_0 - [A_1]_0)e^{([A_1]_0 - [A_2]_0)k_1 t}H(t)}{[A_2]_0(k_1[A_1]_0 - k_1[A_2]_0 + k_2)} \tag{2.88}$$

$$H(t) = {}_2F_1\left(1, 1 + \frac{k_2}{k_1([A_1]_0 - [A_2]_0)}, 2 + \frac{k_2}{k_1([A_1]_0 - [A_2]_0)}, \frac{[A_1]_0}{[A_2]_0} e^{([A_1]_0 - [A_2]_0)k_1 t}\right)$$

In this formula ${}_2F_1$ means the first hypergeometric function, which is defined as follows:

$$_2F_1(a, b, c, x) = 1 + \sum_{n=1}^{\infty} \frac{a(a+1)\cdots(a+n-1)b(b+1)\cdots(b+n-1)}{c(c+1)\cdots(c+n-1)} \frac{x^v}{n!} \tag{2.89}$$

If $[A_1]_0 = [A_2]_0$, the solution can be obtained by the previously stated case of the second order-first order process (Eq. (2.85)). In the case where

$k_2 = jk_1([A_1]_0 - [A_2]_0)$ is true (where i is a positive integer), special solutions arise. For $j = 1$, the solution takes the following form:

$$[A_3]_t = [A_3]_0 e^{-k_2 t} + e^{-k_2 t}[A_1]_0 \left(1 - \frac{[A_1]_0}{[A_2]_0}\right) \ln \frac{[A_2]_0 e^{k_2 t} - [A_1]_0}{[A_2]_0 - [A_1]_0} +$$

$$\frac{[A_1]_0^2(1 - e^{k_2 t})}{[A_2]_0 e^{k_2 t} - [A_1]_0}$$

$$(2.90)$$

For $j > 1$, there are also special forms for the solution that can be obtained in an analogous manner.

A logical next possibility is a third order reaction followed by a first order process, which is represented as follows [10]:

$$3A_1 \xrightarrow{k_1} A_2 \xrightarrow{k_2} A_3 \qquad (2.91)$$

The concentration of A_1 is obtained easily from a single-concentration rate law as in the previous cases, whereas the concentration of A_2 is given by the following expression:

$$[A_2]_t = [A_2]_0 e^{-k_2 t} + \frac{[A_1]_0}{3} e^{-k_2 t} + \frac{[A_1]_0}{3\sqrt{1 + 6k_1[A_1]_0^2 t}}$$

$$- \sqrt{\frac{\pi k_2}{54 k_1}} e^{-k_2 t - k_2/(6k_1[A_1]_0^2)} \mathrm{erfi}\left(\sqrt{\frac{k_2}{6k_1[A_1]_0^2}}\right)$$

$$+ \sqrt{\frac{\pi k_2}{54 k_1}} e^{-k_2 t - k_2/(6k_1[A_1]_0^2)} \mathrm{erfi}\left(\sqrt{\frac{k_2(1 + 6k_1[A_1]_0^2 t)}{6k_1[A_1]_0^2}}\right)$$

$$(2.92)$$

In this expression, $\mathrm{erfi}(x)$ refers to the imaginary error function, which is most easily defined through the complex error function $\mathrm{erf}(x)$, using the imaginary unit $i = \sqrt{-1}$ as follows:

$$\mathrm{erfi}(x) = -i \times \mathrm{erf}(i \times x) \qquad (2.93)$$

The next consecutive process for which the analytical solution can be given is composed of a first order reaction followed by a second order process:

$$A_1 \xrightarrow{k_1} A_2$$
$$2A_2 \xrightarrow{k_2} A_3 \qquad (2.94)$$

Again, the concentration of A_1 can be determined from a single-concentration rate law and follows single exponential decay. $[A_2]$ is given as:

$$[A_2]_t = \sqrt{\frac{k_1[A_1]_0 e^{-k_1 t}}{2k_2} \frac{2K_1\left(\sqrt{8k_2[A_1]_0 e^{-k_1 t}/k_1}\right) - \omega I_1\left(\sqrt{8k_2[A_1]_0 e^{-k_1 t}/k_1}\right)}{2K_0\left(\sqrt{8k_2[A_1]_0 e^{-k_1 t}/k_1}\right) + \omega I_0\left(\sqrt{8k_2[A_1]_0 e^{-k_1 t}/k_1}\right)}}$$

$$\omega = \frac{\sqrt{8k_1[A_1]_0/k_2}\, K_1\left(\sqrt{8k_2[A_1]_0/k_1}\right) - 4[A_2]_0 K_0\left(\sqrt{8k_2[A_1]_0/k_1}\right)}{\sqrt{2k_1[A_1]_0/k_2}\, I_1\left(\sqrt{8k_2[A_1]_0/k_1}\right) + 2[A_2]_0 I_0\left(\sqrt{8k_2[A_1]_0/k_1}\right)} \qquad (2.95)$$

Here, I_1 and I_0 are the modified Bessel functions of the first kind, K_1 and K_0 are the modified Bessel functions of the second kind.

$$I_n(x) = \left(\frac{x}{2}\right)^n \sum_{j=0}^{\infty} \frac{1}{j!(n+j)!} \left(\frac{x}{2}\right)^j$$

$$K_n(x) = x^n \prod_{j=1}^{n}(2j-1) \int_0^\infty \frac{\cos w}{(w^2 + x^2)^{n+1/2}} dw \qquad (2.96)$$

Finally, the case of a second order reaction followed by a second order process can also be handled in this way:

$$2A_1 \xrightarrow{k_1} A_2$$
$$2A_2 \xrightarrow{k_2} A_3 \qquad (2.97)$$

The concentration of A_1 follows second order decay in this scheme. The concentration of A_2 is obtained as follows:

$$[A_2]_t = \frac{k_1[A_1]_0}{2k_2} \frac{(1+\xi)(\sqrt{1+2k_1[A_1]_0 t})^{-1+\xi} + \omega(1-\xi)(\sqrt{1+2k_1[A_1]_0 t})^{-1-\xi}}{(\sqrt{1+2k_1[A_1]_0 t})^{1+\xi} + \omega(\sqrt{1+2k_1[A_1]_0 t})^{1-\xi}}$$

$$\xi = \sqrt{1+2k_2/k_1} \qquad \omega = \frac{k_1(1+\xi) - 2k_2[A_2]_0/[A_1]_0}{2k_2[A_2]_0/[A_1]_0 + k_1(\xi-1)} \qquad (2.98)$$

Figure 2.6 displays some example curves for the concentration of the intermediate in the schemes discussed in the previous paragraphs. The parameters of all of the shown curves were selected so that the initial rate of the formation of A_2 is the same.

Among the single-concentration rate equations, the one describing isotope exchange in Eq. (2.38) might have been the most unusual. This process can be

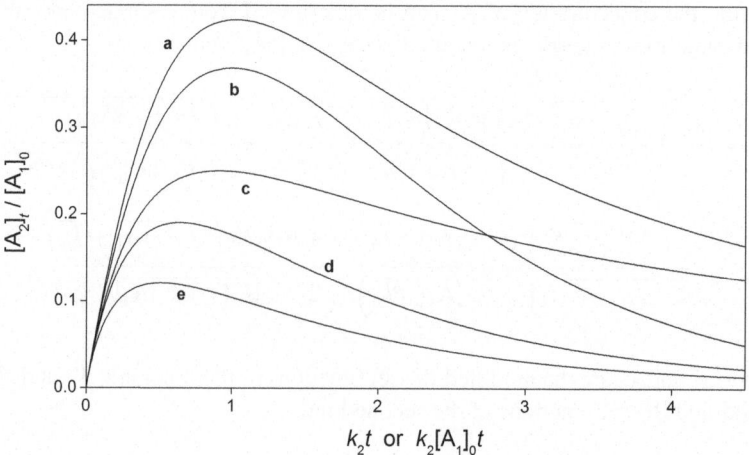

Fig. 2.6 Scaled kinetic traces for consecutive reactions. Formulas used: Eq. (2.95) $k_1/(k_2[A_1]_0) = 1$ (*a*), Eq. (2.98) $k_1/k_2 = 1$ (*b*), Eq. (2.60) $k_1/k_2 = 1$ (*c*), Eq. (2.85) $k_1[A_1]_0/k_2 = 1$ (*d*), Eq. (2.92) $k_1[A_1]_0^2/k_2 = 1$ (*e*)

coupled to first order decays in the number of labels, which gives rise to the following scheme:

$$A_1^* + A_2 \rightleftharpoons A_1 + A_2^*$$

$$A_1^* \xrightarrow{k_1} A_1 \tag{2.99}$$

$$A_2^* \xrightarrow{k_2} A_2$$

This seemingly artificial scheme actually has immense importance in the field of medical diagnostics, and it also forms the basis of measuring exchange rates in nuclear magnetic resonance spectroscopy. The labeling in these cases, rather than isotope substitution in Eq. (2.38), arises from the magnetization of nuclei. The quantities x_1 and x_2 are defined as given in Eq. (2.39) earlier. The differential equations describing the change in these abundances are as follows:

$$\frac{dx_1}{dt}[A_1]_T = -(R_{ex} + k_1[A_1]_T)x_1 + R_{ex}x_2$$

$$\frac{dx_2}{dt}[A_2]_T = -(R_{ex} + k_2[A_2]_T)x_2 + R_{ex}x_1 \tag{2.100}$$

The solution of this system of linear homogeneous differential equations is given by double exponential functions as follows:

$$x_{1,t} = \left(\frac{(x_{2,0}-x_{1,0})R_{ex}}{(\lambda_1-\lambda_2)[A_1]_T} + x_{1,0}\frac{\lambda_2+k_1}{\lambda_2-\lambda_1} \right) e^{\lambda_1 t}$$

$$+ \left(\frac{(x_{2,0}-x_{1,0})R_{ex}}{(\lambda_2-\lambda_1)[A_1]_T} + x_{1,0}\frac{\lambda_1+k_1}{\lambda_1-\lambda_2} \right) e^{\lambda_2 t} \tag{2.101}$$

$$x_{2,t} = \left(\frac{(x_{1,0}-x_{2,0})R_{ex}}{(\lambda_1-\lambda_2)[A_2]_T} + x_{2,0}\frac{\lambda_2+k_2}{\lambda_2-\lambda_1} \right) e^{\lambda_1 t}$$

$$+ \left(\frac{(x_{1,0}-x_{2,0})R_{ex}}{(\lambda_2-\lambda_1)[A_2]_T} + x_{2,0}\frac{\lambda_1+k_2}{\lambda_1-\lambda_2} \right) e^{\lambda_2 t} \tag{2.102}$$

The values λ_1 and λ_2 are the eigenvalues of a matrix similar to $\underline{k}$ given in Eq. (2.76):

$$\lambda_1 = -\left(\frac{k_1}{2} + \frac{k_2}{2} + \frac{R_{ex}}{2[A_1]_T} + \frac{R_{ex}}{2[A_2]_T} \right)$$

$$+ \sqrt{\left(\frac{k_1}{2} + \frac{k_2}{2} + \frac{R_{ex}}{2[A_1]_T} + \frac{R_{ex}}{2[A_2]_T} \right)^2 - k_1 k_2 - \frac{k_1 R_{ex}}{[A_2]_T} - \frac{k_2 R_{ex}}{[A_1]_T}}$$

$$\lambda_2 = -\left(\frac{k_1}{2} + \frac{k_2}{2} + \frac{R_{ex}}{2[A_1]_T} + \frac{R_{ex}}{2[A_2]_T} \right) \tag{2.103}$$

$$- \sqrt{\left(\frac{k_1}{2} + \frac{k_2}{2} + \frac{R_{ex}}{2[A_1]_T} + \frac{R_{ex}}{2[A_2]_T} \right)^2 - k_1 k_2 - \frac{k_1 R_{ex}}{[A_2]_T} - \frac{k_2 R_{ex}}{[A_1]_T}}$$

The usual significance of these equations is that R_{ex} can be calculated from λ_1 and λ_2, which can be obtained directly from the detected kinetic curves.

As already pointed out, multiple-concentration rate equations often do not have known analytical solutions. Yet, it is difficult to predict the cases in which the solution can actually be given and some surprises may await the researchers. Usually, manual attempts at solving a rate equation are very time-consuming, but in all cases, trying to find the analytical solution using the differential equation solver of a symbolic software such as Mathematica is usually short, and may actually find the solution for most of the cases where it is available.

In multiple-concentration rate equations, it is possible that an explicit solution as a function of time cannot be given for the concentrations, yet an explicit formula connecting the concentrations to each other can be obtained. If concentration data are available for these components, this limited analytical formula provides some possibility to test adherence to the assumed rate equation. Two examples will be given in the following paragraphs.

The first example is the scheme where a reactant is transformed into a product in a first order step, then this product reacts with a second molecule of the reactant in a second order step. The scheme is given as follows:

$$A_1 \xrightarrow{k_1} A_2$$

$$A_1 + A_2 \xrightarrow{k_2} A_3$$

(2.104)

The corresponding rate equation is:

$$\frac{d[A_1]}{dt} = -k_1[A_1] - k_2[A_1][A_2]$$

$$\frac{d[A_2]}{dt} = k_1[A_1] - k_2[A_1][A_2]$$

(2.105)

As the concentration of component A_1 decreases monotonously, it makes sense to seek a function that describes the concentration of A_2 as function of the concentration of A_1 and not time. The concentration of A_2 changes monotonously: Eq. (2.105) shows that $[A_2] > k_1/k_2$ would be needed for a decrease, yet the initial concentration is zero, which guarantees that this state can never be reached. The mathematical background of the technique employed here is the derivation of composite functions, but an often useful shortcut is simply to think that two lines of Eq. (2.105) are "divided" by each other (see the footnote for the paragraph before Eq. (2.1)). The result is a differential equation without time derivatives:

$$\frac{d[A_1]}{d[A_2]} = \frac{k_1 + k_2[A_2]}{k_2[A_2] - k_1}$$

(2.106)

The concentration of component A_1 does not appear on the right-hand side of Eq. (2.106), so it is clearly a separable differential equation, which can be solved readily. The typical case is when $[A_2]_0 = 0$, for which the solution is:

$$[A_1]_t = [A_1]_0 + [A_2]_t + \frac{2k_1}{k_2} \ln\left(1 - \frac{k_2[A_2]_t}{k_1}\right)$$

(2.107)

This formula connects the concentrations of A_1 and A_2, which might be useful even without an analytical formula for the time dependence. This technique might also aid finding an analytical solution, as substitution of the formula gained here into the original rate equation will lead to a separable single concentration rate equation.

The second example is a competition reaction, where two reactants react with a common material in a pair of second order processes:

$$A_1 + A_3 \xrightarrow{k_1} A_4$$
$$A_2 + A_3 \xrightarrow{k_2} A_5$$

(2.108)

The rate equation is given as follows:

$$\frac{d[A_1]}{dt} = -k_1[A_1][A_3]$$

$$\frac{d[A_2]}{dt} = -k_2[A_2][A_3]$$

(2.109)

Although further equations would describe the rates of change for the other components, these two suffice for our present purposes. Both $[A_1]$ and $[A_2]$ decrease monotonously. Therefore, they can be described by changing the independent variable from time to a concentration. The resulting differential equation is:

$$\frac{d[A_2]}{d[A_1]} = \frac{k_2[A_2]}{k_1[A_1]}$$

(2.110)

Again, the solution this separable differential equation is readily obtained:

$$[A_2]_t = [A_2]_0 \left(\frac{[A_1]_t}{[A_1]_0} \right)^{\frac{k_2}{k_1}}$$

(2.111)

Curves in Fig. 2.7 show the two concentrations as a function of each other. The first model has one scaling parameter ($[A_1]_0$) and one shape parameter ($k_2[A_1]_0/k_1$) as the typical initial concentration of A_2 is zero. In fact, it could be shown with two scaling parameters and no shape parameters, but for this, $[A_1] - [A_2]$ would need

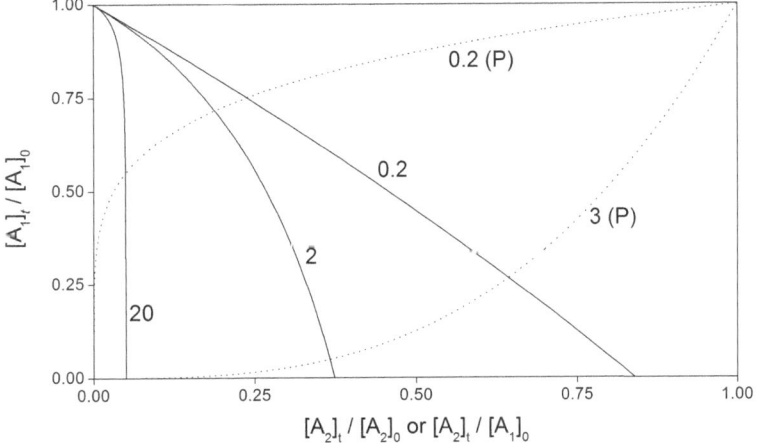

Fig. 2.7 Scaled concentrations as a function of each other in two mechanisms. *Solid lines*: Eq. (2.104) with $k_2[A_1]_0/k_1$ as the shape parameter. *Dotted lines* (also marked with P): Eq. (2.108) with k_1/k_2 as the shape parameter

to be plotted as function of $[A_2]$, which would make an unusual representation, the utility of which is not obvious. The second model has two scaling parameters ($[A_1]_0$ and $[A_0]_0$) and one shape parameter (k_1/k_2).

2.2 Numerical Methods

A **numerical solution** of a rate equation usually gives a collection of numbers rather than functions. The process is also called **numerical integration**, which generates approximate values of concentrations at predetermined time intervals.

The simplest numerical integration method is called the **Euler method**, in which the product of a preselected small time step (τ) and the derivative given in the rate equation is simply added to the concentration of A_i to obtain the new concentration value at the later time:

$$[A_i]_{t+\tau} = [A_i]_t + f_i([A_1]_t, [A_2]_t, \ldots, [A_n]_t)\tau \tag{2.112}$$

The additive term $f_i([A_1]_t, [A_2]_t, \ldots, [A_n]_t)\tau$ on the right is called a **concentration increment**. This method is applied in a progressive manner, giving approximate concentrations at gradually increasing times.

A more advanced method is called the fourth order **Runge–Kutta method** [7, 17]. The term "fourth order" refers to the fact that four different approximations of the concentration increment are estimated with the following formulas:

$$\begin{aligned}
h_{i,1} &= f_i([A_1]_t, [A_2]_t, \ldots, [A_n]_t) \\
h_{i,2} &= f_i([A_1]_t + h_{1,1}\tau/2, [A_2]_t + h_{2,1}\tau/2, \ldots, [A_n]_t + h_{n,1}\tau/2) \\
h_{i,3} &= f_i([A_1]_t + h_{1,2}\tau/2, [A_2]_t + h_{2,2}\tau/2, \ldots, [A_n]_t + h_{n,2}\tau/2) \\
h_{i,4} &= f_i([A_1]_t + h_{1,3}\tau, [A_2]_t + h_{2,3}\tau, \ldots, [A_n]_t + h_{n,3}\tau)
\end{aligned} \tag{2.113}$$

Finally, the concentration increment actually used for a given species is calculated as the weighted average of the four different approximated terms.

$$[A_i]_{t+\tau} = [A_i]_t + \frac{h_{i,1} + 2h_{i,2} + 2h_{i,3} + h_{i,4}}{6}\tau \tag{2.114}$$

A still more advanced method is dependent of the use of Taylor series. According to the Taylor theorem, the concentration $[A_i]_{t+\tau}$ can be calculated from concentration $[A_i]_t$ using the derivatives at time t by the following infinite sum:

$$[A_i]_{t+\tau} = [A_i]_t + \sum_{j=1}^{\infty} \frac{1}{j!} \frac{d^j[A_i]}{dt^j} \tau^j \tag{2.115}$$

The first derivative $(j = 1)$ is already known from the rate equation to be $f_i([A_1], [A_2], \ldots, [A_n])$. The higher order derivatives can be calculated from the rate equation by successive differentiation:

$$\frac{d^j [A_i]}{dt^j} = \sum_{k=1}^{n} \frac{\partial \frac{d[A_k]^{j-1}}{dt^{j-1}}}{\partial [A_k]} \frac{d[A_k]}{dt} = \sum_{k=1}^{n} \frac{\partial \frac{d[A_k]^{j-1}}{dt^{j-1}}}{\partial [A_k]} f_k([A_1], [A_2], \ldots, [A_n])$$

(2.116)

The sum is calculated to a finite number of terms in the series. The higher the number of terms considered, the better the approximation becomes. It should be noted that the Euler method is the same as the Taylor series method truncated to a single additive term (i.e., stopping at the first derivative).

The Taylor series method can sometimes provide analytical solutions as well if the resulting infinite sum is recognized as the exact Taylor series of a particular function. For the simple first order rate equation, this is a very spectacular way of proving that the solution is the exponential function. In this case, $f_1([A_1]) = -k[A_1]$, and the derivatives at $t = 0$ can all be calculated quite simply:

$$\left(\frac{d^j [A_1]}{dt^j} \right)_0 = (-k)^j [A_1]_0$$

(2.117)

Substituting these derivatives into Eq. (2.115) with $t = 0$ gives the following formula:

$$[A_i]_\tau = [A_i]_0 + \sum_{j=1}^{\infty} \frac{1}{j!} (-k)^j [A_1]_0 \tau^j = [A_1]_0 \sum_{j=0}^{\infty} \frac{(-k\tau)^j}{j!}$$

(2.118)

The infinite sum on the right of this equation is exactly the value of e^{-kt} by the definition of the exponential function.[2]

The key question in using numerical integration is the selection of suitable τ time step values. Too small a value requires a lot of computational power, whereas the numerical calculations do not approximate the solution well if τ is too high. In addition, it often happens that at the beginning of the process, very low values of τ are needed, but later on, the time steps can be increased significantly without any loss in accuracy. A numerical instability arising from this sort of sensitivity to the time step is called **stiffness** [2]. When stiff differential equations are solved numerically, it is imperative to use variable time steps. The **Gear algorithm**, also called **predictor-corrector method** [5] provides a common way of solving stiff systems in chemical kinetics.

---

[2]Hungarian-born mathematician Pál Erdős often told his audience that God has a book for recording the most elegant proofs of all theorems. As he used to say, mathematicians need not believe in God, but all of them must believe in the book. The author is sure this proof is from God's book.

In all of these numerical methods, there are two conceptually different sources of error, which cause the differences between the results of the numerical calculations and the actual solution. The first is called **formula error**, which originates from the approximate nature of the formulas used to calculate the concentration increment. The second kind of error is called **propagating error**, which is caused by the fact that the value $[A_i]_t$ used in calculating the next value $[A_i]_{t+\tau}$ is imprecise except in the case of $t = 0$.

There are a number of different softwares that can solve almost any rate equation numerically. Here is a list of softwares this author has some familiarity with: COPASI [1], KINSIM [6], Pro-Kineticist [15], Scientist [14], SPECFIT [18], ZiTa [19]. All of these softwares can solve kinetic differential equations numerically, some of them even contain a least squares minimizing algorithm that is suitable for finding the combination of parameters that best fit a set of experimentally measured data.

2.3   Fitting to Measured Data

Fitting to measured data is a very common problem in chemical kinetics. It serves two different purposes, the more important of which is to test whether detected data follow some sort of theoretical function or not. The second purpose is to determine the values of the parameters appearing in the theoretical function, which are often used to draw further conclusions. It should be emphasized that checking the quality of the fit takes precedence over determining the parameter values. The usual mathematical procedures used in fitting yield parameter values even if the measured data do not resemble the theoretical function at all. However, if the quality of the fit is not sufficient, the determined parameters do not contain any physical information. Unfortunately, this sort of **force fitting**, i.e., using theoretical functions that do not describe the measured data well is very common today's science (e.g., exponential curves are used to fit kinetic traces even if the curves are spectacularly far away from being exponential). This is often caused by an insufficient knowledge of possible theoretical functions describing the phenomenon and the belief that one of a very narrow selection of functions must describe the results for theoretical reasons. It cannot be emphasized strongly enough that no scientific conclusion can be drawn from the values of such force-fitted parameters. The primary objective of fitting is to learn whether the theoretical function describes the data well, and the determined parameter values only make sense if this first question is answered positively.

There are a huge number of scientific softwares that use built-in algorithms to carry out fitting and statistical analysis. These algorithms could even be used as a black box (without any knowledge of the principles on which they are based), but for scientific users, it is highly advisable to get familiar at least with the basics of the mathematical background. It is also a good idea to test computer softwares used for such statistical analysis before first use and compare their results against some sort of standard, well-known data set.

The most common mathematical method used in fitting is called least squares fitting. The term "least squares" refers to the fact that those parameter values are determined for which the square of the difference between the measured points and the theoretical function is minimal. The procedure is also often called nonlinear, which should be interpreted not in the context of the independent variable, but with respect to the parameters. For example, consider the following second order polynomial:

$$g(t) = C_1 t^2 + C_2 t + C_3 \tag{2.119}$$

This is a nonlinear function with respect to the independent variable t, but it is linear for the parameters C_1, C_2, and C_3.

Let $g(t)$ be the theoretical function of interest, N the number of measured points, $g_1, g_2, \ldots, g_N$ the values of the dependent variable experimentally determined for the independent variables $t_1, t_2, \ldots, t_N$. Some of the values $t_1, t_2, \ldots, t_N$ may be equal to each other, this means that the dependent variable was measured multiple times at a certain independent variable.

The difference $g_i - g(t_i)$ is called a **residual**. The sum of the squares of the residuals is calculated and minimized in the fitting procedure. This sum is handled as a function of the parameter values that must be determined. If function $g(t)$ has m different parameters, the sum of squares is defined as follows:

$$S(C_1, C_2, \ldots, C_m) = \sum_{i=0}^{N} (g_i - g(t_i))^2 \tag{2.120}$$

In the process of fitting, those parameter values are sought for which the function $S(C_1, C_2, \ldots, C_m)$ is at minimum. The process is also called **optimization**. From time to time, the physically meaningful values of the parameters may be limited, these limitations can be built into the optimization procedure.

In real-life applications, the different measured points may not be equally important or not equally reliable (these concepts are usually connected). These differences can be taken into account in the fitting procedure by defining weights $(w_1, w_2, \ldots, w_N)$ for each measured point. In this case, the function to be minimized involves multiplication of each residual with its individual weight.

$$S(C_1, C_2, \ldots, C_m) = \sum_{i=0}^{N} w_i (g_i - g(t_i))^2 \tag{2.121}$$

If all weights are equal, that will lead to the same result as no weighting at all $(w_1 = w_2 = \cdots = w_N = 1)$ as given for the function in Eq. (2.120). This is called **uniform weighting**. Uniform weighting assumes that the absolute errors of the measured points are constant, so their reliability is independent of their value.

Another common case is when the relative errors of measured points are constant, i.e., their reliability is directly proportional to their value. The weights for this case, called **proportional weighting**, are the squares of the reciprocals of the measured values ($w_i = 1/g_i^2$).

If the individual standard deviations of all measured points $(\sigma_1, \sigma_2, \ldots, \sigma_N)$ are known, the squares of their inverse reciprocals are best used as weights ($w_i = 1/\sigma_i^2$). This basically means that each point is weighted according to their reliability. However, it is not typical to measure experimental points together with their standard deviations.

Kinetic and rate data are often determined from light absorption measurements. Absorbance is defined as the logarithm of the ratio of two light intensities (one measured in the sample and a reference value). The detectors used in instruments used for measurements (spectrophotometers) typically have an error in the measurement of light intensity that is close to independent of the light intensity. High absorbance values mean low intensity, which is not measured reliably. One solution of this problem is to exclude the use of absorbance values above a cut-off limit (most often absorbance 2, which means that 99 % of the incoming light is absorbed, so 1 % reaches the detector), and use uniform weighting for lower values. Another possibility, which is by no means in widespread use, is to employ special weights based on the assumption of uniform error in intensity measurements. This might be referred to as **absorbance weighting** and takes the following mathematical following form:

$$w_i = 10^{-2g_i} \tag{2.122}$$

Whatever procedure is used, the weights in such a procedure should be defined independently of the results of the fitting. The reliability or importance of the individual data points should not be assessed based on how well they fit to the theoretical equation. Unfortunately, weighting in a way that gives the best fit to the theoretical equation is not unknown in practice. However, this is a highly prejudiced procedure that is a mere mathematical trick to make fitting statistics look good. In fact, this sort of weighting is equivalent to gradually erasing the experimental points that do not fit the theory.

The sum of squares shown in Eqs. (2.120) and (2.121) minimize the difference along the dependent variable and, taken literally, they assume that the values of the independent variable are known with high precision. Of course, this is a useful assumption when the error in the independent variable is at least an order of magnitude lower than the error in the dependent variables. Should this not be the case, different definitions of the sum of squares function may be employed. Statistical procedures based on this idea are often called **Deming regressions** [3]. A special case is **orthogonal fitting**, which minimizes the sum of squared perpendicular distances from the data points to the theoretical function. In chemical kinetics, it is seldom necessary to use these techniques, but a careful experimenter should know about their existence and recognize the special cases when they are needed.

Function $S(C_1, C_2, \ldots, C_m)$, as defined in Eq. (2.121), cannot be negative as it involves summing the squares of real numbers. The minimum of this multivariable function is found where the partial derivatives are zero:

$$\frac{\partial S(C_1, C_2, \ldots, C_m)}{\partial C_i} = 0 \tag{2.123}$$

Solving these equations, which are nonlinear if the theoretical function $g(t)$ is nonlinear with respect to at least one of the parameters, gives the optimized values of the parameters. The number of equations in Eq. (2.123) is the same as the number parameters, so, at least in theory, all parameters can be determined from this equation. Quite often, Taylor series expansion is used to transform Eq. (2.123) into a set of linear equations called **normal equations** using the partial derivatives of function $g(t)$ with respect to the parameters. With this technique, Eq. (2.123) is not solved precisely, but the errors caused by the approximations are usually so small that they are without any consequence in experimental science.

An often neglected point is that when the values of the parameters are determined, the reliability of the values must also be assessed. This is done through estimating the standard deviations of the parameters. The usual course of calculations involves the definition of a matrix, $\underline{\underline{M}}$, based on the products of partial derivatives with respect to the parameters as follows:

$$\underline{\underline{M}} = \begin{pmatrix} \sum_{i=1}^{N} w_i \frac{\partial g(t_i)}{\partial C_1} \frac{\partial g(t_i)}{\partial C_1} & \sum_{i=1}^{N} w_i \frac{\partial g(t_i)}{\partial C_1} \frac{\partial g(t_i)}{\partial C_2} & \cdots & \sum_{i=1}^{N} w_i \frac{\partial g(t_i)}{\partial C_1} \frac{\partial g(t_i)}{\partial C_m} \\ \sum_{i=1}^{N} w_i \frac{\partial g(t_i)}{\partial C_2} \frac{\partial g(t_i)}{\partial C_1} & \sum_{i=1}^{N} w_i \frac{\partial g(t_i)}{\partial C_2} \frac{\partial g(t_i)}{\partial C_2} & \cdots & \sum_{i=1}^{N} w_i \frac{\partial g(t_i)}{\partial C_2} \frac{\partial g(t_i)}{\partial C_m} \\ \vdots & \vdots & \ddots & \vdots \\ \sum_{i=1}^{N} w_i \frac{\partial g(t_i)}{\partial C_m} \frac{\partial g(t_i)}{\partial C_1} & \sum_{i=1}^{N} w_i \frac{\partial g(t_i)}{\partial C_m} \frac{\partial g(t_i)}{\partial C_2} & \cdots & \sum_{i=1}^{N} w_i \frac{\partial g(t_i)}{\partial C_m} \frac{\partial g(t_i)}{\partial C_m} \end{pmatrix} \tag{2.124}$$

This matrix appears to involve a lot of calculations. This is indeed true, but these are highly routine and in a way simple as they only require the knowledge of the theoretical function and the values of t_i. Neither the experimentally measured values g_i nor the optimized values of the parameters are necessary.

As a next step, the inverse of matrix $\underline{\underline{M}}$ needs to be calculated. The elements of this inverse matrix, denoted $\langle \underline{\underline{M}}^{-1} \rangle_{i,j}$ for row i and column j, are essential in calculating further statistical descriptors. The standard deviation of parameter C_i is given as:

$$\sigma_{C_i} = \sqrt{\frac{\langle \underline{\underline{M}}^{-1} \rangle_{i,i} S(C_1, C_2, \ldots, C_m)}{N - m}} \tag{2.125}$$

Another highly significant statistical descriptor is the correlation coefficient between two parameters, which can be defined for any pairs of parameters as follows:

$$c_{C_i, C_j} = \frac{\langle \underline{\underline{M}}^{-1} \rangle_{i,j}}{\sqrt{\langle \underline{\underline{M}}^{-1} \rangle_{i,i} \langle \underline{\underline{M}}^{-1} \rangle_{j,j}}} \tag{2.126}$$

There are a number of other statistical descriptors (such as coefficient of determination, skewness, curtiosis, etc.) that can be used to characterize the quality of a fit. They rarely have major significance, but from time to time, they may be useful.

With this mathematical background, the following paragraphs will show specific examples of fitting through linear and exponential functions.

Fitting a straight line, which is also called **linear regression**, is identical to fitting a first order polynomial to measured data. This is a very common problem, and in earlier times, a ruler and some good judgment in eyesight were enough to obtain reasonably reliable data. The mathematics behind the fitting procedure was well known even then, and using the full statistical calculations became very easy after the personal computers became available for everyday use. The theoretical function in this case is a linear one with two parameters:

$$g(t) = C_1 t + C_2 \tag{2.127}$$

C_1 is commonly called slope and C_2 is called intercept, which is the value of the function at $t = 0$. Another sort of intercept is when $g(t) = 0$, which is not a direct parameter of the function but can be derived from the intercept and slope as $-C_2/C_1$. The equations that arise from setting the partial derivatives of $S(C_1, C_2, \ldots, C_m)$ to zero are as follows:

$$\frac{\partial \sum_{i=0}^{N} w_i (g_i - C_1 t_i - C_2)^2}{\partial C_1} = -2 \sum_{i=0}^{N} w_i (g_i - C_1 t_i - C_2) t_i = 0$$

$$\frac{\partial \sum_{i=0}^{N} w_i (g_i - C_1 t_i - C_2)^2}{\partial C_2} = -2 \sum_{i=0}^{N} w_i (g_i - C_1 t_i - C_2) = 0 \tag{2.128}$$

This is a system of two simultaneous linear equations. The solution yields the following parameter values:

$$C_1 = \frac{\left(\sum_{i=1}^{N} w_i g_i t_i \right) \left(\sum_{i=1}^{N} w_i \right) - \left(\sum_{i=1}^{N} w_i g_i \right) \left(\sum_{i=1}^{N} w_i t_i \right)}{\left(\sum_{i=1}^{N} w_i t_i^2 \right) \left(\sum_{i=1}^{N} w_i \right) - \left(\sum_{i=1}^{N} w_i t_i \right)^2}$$

$$C_2 = \frac{\left(\sum_{i=1}^{N} w_i t_i^2 \right) \left(\sum_{i=1}^{N} w_i g_i \right) - \left(\sum_{i=1}^{N} w_i t_i \right) \left(\sum_{i=1}^{N} w_i g_i t_i \right)}{\left(\sum_{i=1}^{N} w_i t_i^2 \right) \left(\sum_{i=1}^{N} w_i \right) - \left(\sum_{i=1}^{N} w_i t_i \right)^2} \tag{2.129}$$

The standard errors of the individual parameters are calculated as:

$$\sigma_{C_1} = \sqrt{\frac{\sum_{i=1}^{N} w_i}{\sum_{i=1}^{N} w_i \sum_{i=1}^{N} w_i t_i^2 - \left(\sum_{i=1}^{N} w_i t_i\right)^2}} \sqrt{\frac{\sum_{i=0}^{N} w_i (g_i - C_1 t_i - C_2)^2}{N-2}}$$

(2.130)

$$\sigma_{C_2} = \sqrt{\frac{\sum_{i=1}^{N} w_i t_i^2}{\sum_{i=1}^{N} w_i \sum_{i=1}^{N} w_i t_i^2 - (\sum_{i=1}^{N} w_i t_i)^2}} \sqrt{\frac{\sum_{i=0}^{N} w_i (g_i - C_1 t_i - C_2)^2}{N-2}}$$

The correlation between the slope and intercept is given as:

$$c_{C_1,C_2} = \frac{-\sum_{i=1}^{N} w_i t_i}{\sqrt{\sum_{i=1}^{N} w_i \sum_{i=1}^{N} w_i t_i^2}}$$

(2.131)

In chemical kinetics, the most common problem is fitting an exponential function to measured data. This is a clearly nonlinear problem as one of the three parameters appears in the exponent. The form already given in Eq. (2.10) is used:

$$g(t) = Xe^{-kt} + E$$

(2.132)

The optimized values for the linear parameters X and E are easily found as:

$$X = \frac{N \sum_{i=1}^{N} g_i e^{-kt_i} - (\sum_{i=1}^{N} e^{-kt_i})(\sum_{i=1}^{N} g_i)}{N \sum_{i=1}^{N} e^{-2kt_i} - (\sum_{i=1}^{N} e^{-kt_i})^2}$$

$$E = \frac{(\sum_{i=1}^{N} e^{-2kt_i})(\sum_{i=1}^{N} g_i) - (\sum_{i=1}^{N} e^{-kt_i})(\sum_{i=1}^{N} g_i e^{-kt_i})}{N \sum_{i=1}^{N} e^{-2kt_i} - (\sum_{i=1}^{N} e^{-kt_i})^2}$$

(2.133)

However, for the nonlinear parameter k, the equation remains implicit:

$$\left(N \sum_{i=1}^{N} e^{-2kt_i} - \left(\sum_{i=1}^{N} e^{-kt_i}\right)^2 \right) \sum_{i=1}^{N} g_i t_i e^{-kt_i}$$

$$= \left(N \sum_{i=1}^{N} g_i e^{-kt_i} - \sum_{i=1}^{N} e^{-kt_i} \sum_{i=1}^{N} g_i \right) \sum_{i=1}^{N} t_i e^{-2kt_i}$$

$$+ \left((\sum_{i=1}^{N} e^{-2kt_i})(\sum_{i=1}^{N} g_i) - (\sum_{i=1}^{N} e^{-kt_i})(\sum_{i=1}^{N} g_i e^{-kt_i}) \right) \sum_{i=1}^{N} t_i e^{-kt_i}$$

(2.134)

This equation can be solved with a suitable numerical method. Alternatively, the Taylor series expansion can be used the set up a normal equation instead of Eq. (2.134). Then the standard deviations and the correlations of the parameters can be obtained from Eqs. (2.125) and (2.126).

The exponential function is well suited as a probe function to determine rates of concentration change. For this purpose, a relatively small portion of the experimentally detected trace is fitted. The selected portion should contain the time for which the derivative is sought and should be narrow enough so that the exponential function fits quite well. The rate is then calculated by the following formula:

$$\frac{dg(t)}{dt} = -kXe^{-kt} \qquad (2.135)$$

If the initial rate of the process is determined by this technique, the probe function can be re-parametrized so that the initial rate v_0 (the derivative at $t = 0$) appears in it as parameter:

$$g(t) = \frac{v_0}{k}e^{-kt} + E \qquad (2.136)$$

Finally, a further phenomenon related to parameter correlation should be mentioned here. Consider the following three-parameter equation:

$$g(t) = (C_1 + C_2)t + C_3 \qquad (2.137)$$

In this equation, there is no way of determining C_1 and C_2 independently, only their sum $(C_1 + C_2)$ is available from the fitting. In the results of a fitting procedure, this would be apparent from a correlation coefficient of 1 between C_1 and C_2. The example shown here is quite an obvious one as the correlation is caused by the form of the function. In real life examples, such functions are easily recognized. A lot more difficult to recognize is the case when the parameter correlation is caused by the insufficient range of measured data, which cannot be avoided because of experimental limitations. The phenomenon is detected by large values of standard errors and correlations between the parameters close to 1.

Another reason why a particular parameter may have very large standard error as calculated in the fitting is that it does not influence the calculated values of the theoretical function in the range of independent variables covered by the data. In this case, the large standard error is only calculated for a single parameter rather than a pair of C_i values.

References

1. COPASI: http://www.copasi.org/
2. Curtiss, C.F., Hirschfelder, J.O.: Integration of stiff equations. Proc. Natl. Acad. Sci. USA **38**, 235–243 (1952)
3. Deming, W.E.: Statistical Adjustment of Data. Wiley, New York (1943)
4. Érdi, P., Lente, G.: Stochastic Chemical Kinetics—Theory and (Mostly) Systems Biological Applications. Springer, New York (2014)

5. Gear, C.W.: The automatic integration of ordinary differential equations. Commun. ACM **14**, 176–179 (1971)
6. KINSIM/FITSIM: http://www.biochem.wustl.edu/cflab/message.html
7. Kutta, W.: Beitrag zur näherungweisen Integration totaler Differentialgleichungen. Z. Math. Phys. **46**, 435–453 (1901)
8. Lente, G., Fábián, I.: Effect of dissolved oxygen on the oxidation of dithionate ion. Extremely Unusual Kinetic Traces. Inorg. Chem. **43**, 4019–4025 (2004)
9. Lente, G.: Stochastic mapping of first order reaction networks: a systematic comparison of the stochastic and deterministic kinetic approaches. J. Chem. Phys. **137**, 164101 (2012)
10. Lente, G.: Kinetics of irreversible consecutive processes with first order second steps: analytical solutions. J. Math. Chem. DOI: 10.1007/s10910-015-0477-7 (2015)
11. McKay, H.A.C.: Kinetics of exchange reactions. Nature **142**, 997–998 (1938)
12. McKay, H.A.C.: Kinetics of some exchange reactions of the type RI + I*- RI* + I- in alcoholic solution. J. Am. Chem. Soc. **65**, 702–706 (1943)
13. Michaelis, L., Menten, M.L.: Die kinetik der invertinwirkung. Biochem. Z. **49**, 333–369 (1913)
14. Micromath Scientist: http://www.micromath.com/
15. Pro-KIV Kinetic Analysis and Data Simulation Software: http://www.photophysics.com/software/pro-kiv-kinetic-analysis-data-simulation-software
16. Robinson, A.: Non-standard Analysis. Princeton University Press, Princeton (1974)
17. Runge, C., König, H.: Vorlesungen über Numerisches Rechnen. Springer, Berlin (1924)
18. SPECFIT Global Analysis: http://www.hi-techsci.com/products/specfitglobalanalysis/
19. ZiTa by G. Peintler: http://www.staff.u-szeged.hu/~peintler/enprogs.htm

Chapter 3
Inevitable Approximations

Approximations in solving rate equations are used quite generally in chemical kinetics. These approximations were mainly introduced at a time when the possibilities of scientific computation were very limited and approximations often provided the only way to evaluate data in a meaningful manner. As pointed out in Chap. 2, today's computational resources are quite sufficient for the purposes of chemical kinetics and, as a rule, any reasonable rate equation can be solved reliably with a numerical method without any further approximations. Yet the time-honored approximation techniques did not completely lose their importance because they have significance beyond the simplification of mathematical operations necessary.

The fact that approximation methods were (and are) used with great success is primarily a consequence of the limited information content of experimentally measured data. This limitation may simply be some sort of random error in concentration measurements, but typically goes way beyond this uncertainty. It is often impossible (or at least very impracticable because of the unreasonably high costs involved) to measure the concentrations of all the species appearing in a reaction system. Therefore, data evaluation must be based on those concentrations which are available from measurements. Another set of unavoidable limitations is set by the time resolution of the monitoring method, or the time scale of reaction initiation. An attempt to study a process that has a typical time scale that is shorter than the limit of the time resolution is hopeless even if it is only one reaction step in a more complicated system. Yet kinetic information from other steps might be available in the same system only if the effect of the immeasurably fast step is considered somehow.

Using a solution method for the rate equation without approximations certainly cannot give results that are worse than those obtained by approximations. In this sense, the approximations presented in this chapter have little theoretical relevance today: a better method is already known and widely available. However, the unavoidable experimental limitations may easily create a situation when the use of the approximations does not distort any of the information content of the

© Gábor Lente 2015

G. Lente, *Deterministic Kinetics in Chemistry and Systems Biology*, SpringerBriefs in Molecular Science, DOI 10.1007/978-3-319-15482-4_3 experimentally measured data. So the approximations in fact do not introduce any error in these cases. What's more, it is often the situation that the successful use of approximations provides valuable insight into why certain parameters cannot be determined and may even yield clues about further experiments that should be made to improve the information content of measured data. Therefore, the present author believes that these approximations should still be used despite the fact that they are not needed in a mathematical sense any more: as the wording of the title shows, they are inevitable on some level.

The classic textbook of Jim Espenson [9] provides very detailed guidelines on how the various approximations should be used and what information can be deduced from the results. This chapter will not attempt to give a similarly complete summary. Rather, it will focus on how the limited information content of data available is manifested in the success of the approximation methods.

Yet, the introductory paragraphs of this chapter should also include a warning. All the approximation methods discussed here have a more or less defined region of validity. This often means a limited range of initial concentrations for reactants, sometimes the values of certain rate constant must be in a limited range. In other cases, multiple conditions must be satisfied simultaneously. A careful investigator should always test whether these conditions are met in practice and only use the approximations when appropriate. Unfortunately, some chemists value simplicity higher than scientific precision. These examples should not be followed. As Jim Espenson writes [9]: "…the fact that an approximate solution is simple does not mean that it is correct [7, 10]."

3.1   The Method of Flooding

The method of **flooding** is almost as old as chemical kinetics itself. It means that all reactants of a process except one are used in such a high concentration that the change in their concentration is unnoticeably small. In effect, the time dependence of these components is simplified into a constant function:

$$[A_k]_t = [A_k]_0 (= [A_k]_{\text{flooding}}) \tag{3.1}$$

The reagents used in large amount are called **large excess reagents**, whereas the single species with low initial concentration is the **limiting reagent**, but would be better called **deficiency reagent**, because the common use of the phrase limiting reagent does not require that the other reagents are used in large excess. The basic idea of flooding is to simplify the solution of the rate equation of the process by ensuring that all except one of the concentrations remain unchanged during a single kinetic experiment. A system must be flooded with all reagents that influence the rates of concentration change. Products seldom have such an effect, so it is often unnecessary to consider them in flooding. In the exceptional cases when some of the products influence the rates, adding them to the initial mixture in large excess is a good idea to keep up the flooding conditions.

The core idea of flooding is to maintain time-independent concentrations for as many components as possible. The easiest way to achieve this is to use substances in large excess. However, this is not the only possibility. Alternatively, continuously replenishing a consumed substance might serve the same purpose. This is how buffers work in aqueous solution: they maintain constant hydrogen ion concentration by producing or consuming just the appropriate amounts. Similar buffering is not impossible for other species in other systems as well. However, it must always be verified that the replenishing process is fast enough compared to the investigated reactions. For pH buffers, this property is almost guaranteed by the fact that proton transfer reactions are generally known to be among the fastest in aqueous solutions.

How large an excess of reagents must be used for flooding conditions to prevail? This is a critical question for experiment design. The usual rule of thumb used by many active researchers is that ten times the stoichiometrically necessary amount is about the limit above which the approximation of flooding can be used. Some more information about this question will be presented through an example.

A very typical use of flooding is in the study of a second order reaction between two different reagents, which was called a mixed second order reaction in Eq. (2.26):

$$A_1 + A_2 \xrightarrow{k_1} A_3(+\dots)$$

$$\frac{d[A_1]}{dt} = \frac{d[A_2]}{dt} = -k_1[A_1][A_2] \tag{3.2}$$

If the system is flooded with A_2 so that $[A_2]_0 \gg [A_1]_0$, the concentration of A_2 remains approximately constant in time at $[A_2]_t = [A_2]_0 = [A_2]_{\text{flooding}}$, and it is enough to focus on the concentration change of A_1. Substitution of the flooding condition into the rate equation yields the following simplified differential equation:

$$\frac{d[A_1]}{dt} = -k_1[A_1][A_2]_{\text{flooding}} = -k_{\text{obs}}[A_1] \tag{3.3}$$

A new constant, k_{obs}, called the **pseudo-first order rate constant** or **observed rate constant** was introduced above instead of $k_1[A_2]_{\text{flooding}}$. In the textbook of Jim Espenson [9], the notation k_ψ is used for the same purpose because of the Greek origin of the word "pseudo." The solution of the simplified rate equation is a single exponential function:

$$[A_1]_t = [A_1]_0 e^{-k_{\text{obs}}t} \tag{3.4}$$

The usual course of action in this approach is to set up a number of experiments with different initial concentrations of A_2 (but all of them within the range of flooding). The success of the exponential fit is evidence for the first order dependence with respect to reagent A_1 (hence the fit itself is called **pseudo-first order**), whereas a direct proportionality between determined k_{obs} values with $[A_2]_0$ proves the first order nature with respect to reagent A_2. The second order rate constant k_1 is obtained from the slope of the straight line in the plot of k_{obs} versus $[A_2]_0$.

A careful reader might have already realized that exponential curves are fitted in this technique to experimental curves that should not be exactly exponential, the exact solution is given as Eq. (2.31). So this procedure has some built-in approximation error as the actually detected curves will deviate somewhat from the exponential function used for fitting. It is intuitively clear that this error is smaller as the ratio $[A_2]_0/[A_1]_0$ increases. The theoretical residuals (the difference between actual and exponentially fitted values) for a number of cases with different initial concentration ratios are given as a function of time in Fig. 3.1.

The data in Fig. 3.1 reveal that large residuals occur at the very beginning of these traces, serving as a clue that the wrong function used for fitting. The residuals also show a strong tendency. However, some clear tendency is also present when the excess of A_2 is quite high, ten-fold (curve **e**) or 20-fold (curve **f**). The experimental detectability of this tendency depends on the usual error of the measured data. It is worth remembering that such a tendency in the residuals of exponential fits might arise from the approximation of flooding, and not from a failure of the model.

A fitting will give a final value for the parameters no matter what its quality is. It is sometimes recommended that k_1 should not be calculated as $k_1 = k_{obs}/[A_2]_0$, but $k_1 = k_{obs}/[A_2]_{av}$ must be used, where $[A_2]_{av}$ means the time-average concentration of A_2, which can be approximated as the average of initial and final values, which

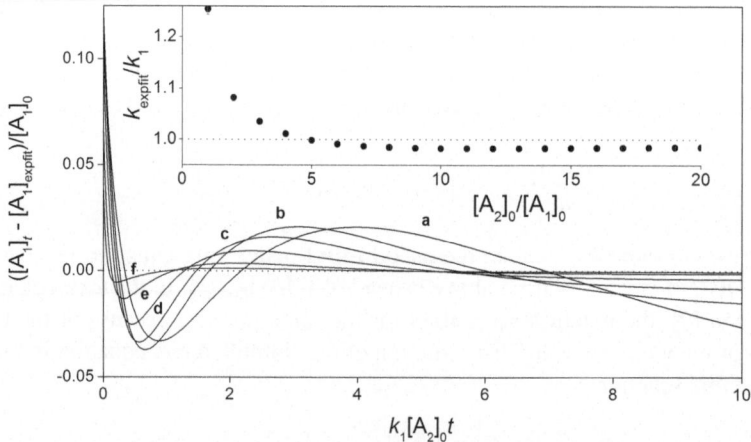

Fig. 3.1 Scaled residuals of exponential fits to a mixed second order reaction (Eq. (2.31)). $[A_2]_0/[A_1]_0 = 1$ (a), 2 (b), 3 (c), 5 (d), 10 (e), 20 (f). Inset: Estimates for k_1 as a function of the initial ratio $[A_2]_0/[A_1]_0$

is $[A_2]_0 - [A_1]_0/2$ for the 1:1 stoichiometry considered here. The inset of Fig. 3.1 shows the results of such an estimate as a function of the initial ratio $[A_2]_0/[A_1]_0$. The value of k_1 estimated by flooding is within 2 % of the precise value whenever the mentioned initial ratio is larger than 4, and the relative difference is not improved much as $[A_2]_0/[A_1]_0$ increases. One might but conclude that a fourfold excess is already enough for the flooding approximation to work well if the sole purpose is the determination of the second order rate constant. The danger in this line of thought is that the model verification aspect of fitting is neglected.

It should also be pointed out that flooding conditions are almost always satisfied toward the end of a kinetic measurement. In consequence, trying to fit the ends of kinetic curves may reveal very useful information even in complicated systems. This procedure typically involves attempts to fit some final parts of kinetic traces with exponential functions and tries to establish the rate equation by noting the dependence of the k_{obs} values on the calculated final concentrations of the other reagents. There is no reason why this practice should be limited to first order dependences. If the process is suspected to be second order with respect to the limiting reagent, fully analogous pseudo-second order evaluation is also possible. However, more caution is needed than in pseudo-first order cases because the calculation of a second order rate constant from instrumental data requires the exact knowledge of the relationship between the signal and the concentrations.

The primary benefit from flooding was originally a simplification of mathematical procedures. With the availability of personal computers, this is not necessary any more. Flooding usually has an increasing effect on rates of concentration change compared to measurements carried out under conditions when the initial concentrations of the reagents are similar. Avoiding the method of flooding is sometimes a useful trick to lengthen reaction times, which may actually make the difference between feasible and unfeasible conditions if the studies are close to the experimental limits of time resolution or mixing time [6, 17].

3.2 Kinetically Separate Phases

Consider the scheme of mixed second order reaction followed by a first order reaction, which was also discussed in Eq. (2.87) of Chap. 2:

$$A_1 + A_2 \xrightarrow{k_1} A_3 \xrightarrow{k_2} A_4 \tag{3.5}$$

The analytical solution of this scheme is presented by Eq. (2.88). Based on this solution, Fig. 3.2 shows some sample curves for the concentration of intermediate A_3 in this scheme for conditions under which the ratio $k_1[A_1]_0/k_2$ changes, but all other parameters remain the same.

The traces in Fig. 3.2 show two different time scales: the first part has $1/(k_1[A_2]_0)$ as a time unit, the second part has $1/k_2$. Species A_3 is clearly a major intermediate in

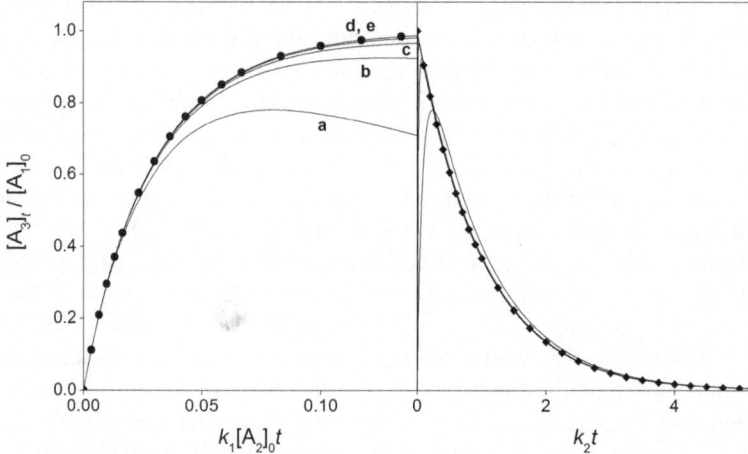

Fig. 3.2 Scaled concentrations as a function of scaled time for intermediate A_3 in the scheme composed of a mixed second order reaction followed by a first order reaction (Eq. (3.5)) at a $[A_2]_0/[A_1]_0$ ratio of 6. $k_1[A_1]_0/k_2 = 2$ (*a*), 10 (*b*), 30 (*c*), 100 (*d*), 1,000 (*e*). Markers represent the traces calculated by assuming kinetic separation

this system, as practically all of the reactants are transformed to A_3 before the onset of later processes. Calculated curves in this scaled representation change very little above the $k_1[A_1]_0/k_2$ ratio of 10. The changes in the shapes of the curves above the ratio of 30 would most certainly be impossible to detect by the usual experimental methods. It is said that at high $k_1[A_1]_0/k_2$ ratios, the two consecutive reactions are **kinetically separated**. Kinetically separated steps can be described independently of each other. In effect, the second step is much slower than the first one, so the first reaction can go to practical completion without any interference from the second. Using this approximation of kinetic separation, the time dependence of the concentration of intermediate A_3 can be given in a much simpler formula than that shown in Eq. (2.88):

$$[A_3] = \frac{[A_2]_0[A_1]_0(1-e^{-k_1([A_2]_0-[A_1]_0)t})}{[A_2]_0-[A_1]_0e^{-k_1([A_2]_0-[A_1]_0)t}} \quad \text{if} \quad t < \frac{1}{k_1[A_2]_0}$$

$$(3.6)$$

$$[A_3] = [A_1]_0e^{-k_2t} \quad \text{if} \quad t \geq \frac{1}{k_1[A_2]_0}$$

The concentrations calculated with these formulas are also given in Fig. 3.2 as markers. It is clear that these approximate concentrations are in very good agreement with the exact solution when the $k_1[A_1]_0/k_2$ ratio is 30 or higher, and usual experimentally measured concentrations would not be able to distinguish between the approximate and the precise solution.

The phenomenon of kinetically separated steps almost always arises when the typical time scale of a later process in a series of consecutive reactions is a lot longer than the time scale of an earlier step. Such steps can be handled independently of each other, so their differential equations can be solved without recognizing the existence of the other. However, the phenomenon also serves as a warning about the evaluation of measured data. A kineticist must constantly ask (and answer!) the following questions: Is there a kinetically separated step that is completed before the first measured point is detected? Did the experiments miss a kinetically separated step simply because the concentration changes were not monitored long enough?

3.3 Rate Determining Steps

Consider the scheme of mixed second order followed by a first order reaction given in Eq. (3.5) again. In the previous section, the cases of interest were mainly those in which the typical time scale of the second process was much longer than that of the first process. In the present considerations, this relation is reversed and the product A_4 will be in the center of the focus instead of the intermediate A_3. Some selected kinetic traces calculated based on Eq. (2.88) are shown in Fig. 3.3. This figure shows scaled kinetic curves for which the only difference is the value of the ratio $k_2/(k_1[A_1]_0)$ (this is the reciprocal of the ratio used in the previous section, the rationale for this change is purely esthetic).

Figure 3.3 clearly highlights that kinetic curves become close to independent of the $k_2/(k_1[A_1]_0)$ ratio when its value exceeds 100. This ratio means that the typical time scale of the second (first order) process is much shorter than the time scale of the first (second order process). These relative time scales ensure that whenever

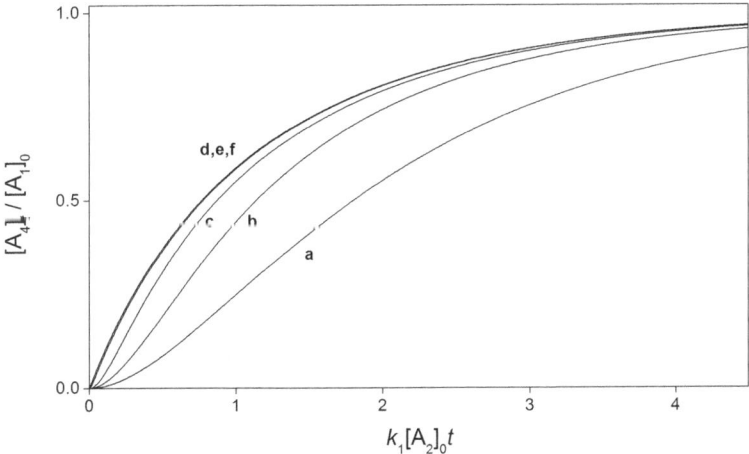

Fig. 3.3 Scaled concentrations as a function of scaled time for product A_4 in the scheme composed of a mixed second order reaction followed by a first order reaction (Eq. (3.5)) at a $[A_2]_0/[A_1]_0$ ratio of 6. $k_2/(k_1[A_1]_0) = 1$ (a), 3 (b), 10 (c), 100 (d) 100,000 (e). Curve **f** was calculated form the scheme of a mixed second order reaction without considering the intermediate (Eq. (3.7))

particles of species A_3 form, they are very rapidly transformed to final product A_4, so species A_3 is an example of a minor intermediate in this system. Under these conditions, the first step is the **rate limiting** or **rate determining** step of the scheme (sometimes rate constant k_1 is also called rate determining). Because species A_3 does not build up to high concentrations, the concentration of the final product A_4 can be calculated simply as a direct product of the first step, which is a mixed second order process:

$$[A_4] = \frac{[A_2]_0[A_1]_0(1 - e^{-k_1([A_2]_0-[A_1]_0)t})}{[A_2]_0 - [A_1]_0 e^{-k_1([A_2]_0-[A_1]_0)t}} \tag{3.7}$$

Curve **f** in Fig. 3.3 was calculated by this formula. It is indistinguishable from curves **d** and **e**, which were calculated for $k_2/(k_1[A_1]_0)$ ratios of 100 and 100,000, respectively. Again, the difference between these three curves is really small, way below the usual level error of experimental methods.

The value of k_2 does not even appear in Eq. (3.7), so, maybe somewhat surprisingly, the calculations of the time-dependent concentration of product A_4 does not require the knowledge of the rate constant of the process that produces it. This is why k_1 is termed rate determining: it sets the pace of the concentration change of the final product. In general, even rather long sequences of reactions may be easy to characterize kinetically if the rate determining step occurs early.

One cannot hope to determine the rate constants of steps after the rate determining step (at least not from observing the buildup of the final product or the loss of reagents). In a few examples, however, the ratios of rate constants of parallel reactions occurring after the rate determining step but leading to different products can be resolved. Strictly speaking ratios like this do not give kinetic information. It is also noticeable that such ratios are often dimensionless, or at the very least, time does not appear in their physical dimensions.

3.4 Steady State Approximation

Again we consider the scheme given in Eq. (3.5) under the general conditions employed in the previous section, but instead of the concentration of the product A_4, the concentration of the intermediate A_3 will be of interest once more. We note that the approximation using the concept of the rate limiting step does not give any meaningful concentrations for this intermediate (or, to make a little too definitive of a statement, it says that the concentration of A_3 is zero during the entire process). The key to this contradictory state of affairs is that A_3 is a minor intermediate, which is unlikely to be detected by experimental methods. Therefore, knowledge of its concentration is not particularly useful.

If the concentration of a minor intermediate is needed, using the **steady state approach** is typically an approximation that works reasonably well. This approach is occasionally also called the **Bodenstein principle** in honor of the scientist who seems to have been the first to describe it in the chemical literature [4].

Jim Espenson's textbook [9] offers a very detailed and highly authoritative description of the use of the steady state approach, it makes no sense to replicate it here, only the basic principles and the usual mathematical handling will be presented with a focus on the principles of use and the limitations.

In a mathematical sense, the steady state approximation means that one of the rates of concentration changes (necessarily that of a minor intermediate) is set to zero at all times. In the Scheme of Eq. (3.5), this is done for minor intermediate A_3 as follows:

$$\frac{d[A_3]}{dt} = k_1[A_1][A_2] - k_2[A_3] = 0 \qquad (3.8)$$

This is not a differential equation any more, and one of the concentrations appearing in it can be expressed as a function of the other two. For the meaningful use of the steady state approximation, it is important that the concentration of the minor intermediate should be given as a function of other concentrations. The chosen intermediate is called a **steady state intermediate**, the expressed concentration is usually referred to as a **steady state concentration** and indicated by the letters "ss" in the subscript:

$$[A_3]_{ss} = \frac{k_1}{k_2}[A_1]_t[A_2]_t \qquad (3.9)$$

In this equation, the time dependence of the other concentrations was deliberately emphasized. Somewhat paradoxically, assuming that the concentration of a minor intermediate does not change in time led to a formula that clearly describes the time-dependent concentration of that intermediate! This paradox will be further analyzed later, but for the time being, the attention should be focused on the usefulness of the formula. Figure 3.4 presents a comparison between precisely calculated concentrations of A_3 and those obtained with the steady state formula under some different conditions.

The comparisons displayed in Fig. 3.4 show that Eq. (3.9) gives a visibly good approximation of the intermediate concentrations after an initial period that is short compared to the time scale of product formation. The notable agreement itself is usually of little importance, as the concentration of a minor intermediate is not typically available from experimental data. However, the steady state concentration expressed in Eq. (3.9) can be used instead of the precise time-dependent concentration in rate equations. For example, the rate of product (A_4) formation in the scheme of Eq. (3.5) can be written as follows:

$$\frac{d[A_4]}{dt} = k_2[A_3]_{ss} = k_2\frac{k_1}{k_2}[A_1]_t[A_2]_t = k_1[A_1][A_2] \qquad (3.10)$$

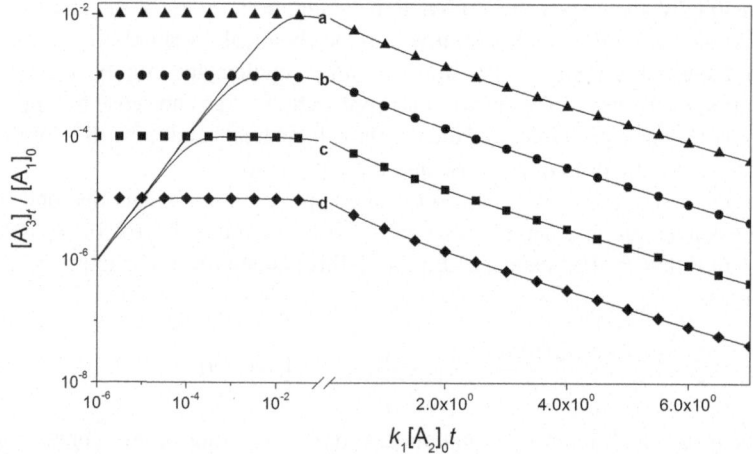

Fig. 3.4 Scaled concentrations as a function of scaled time for intermediate A_3 in the scheme composed of a mixed second order reaction followed by a first order reaction (Eq. (3.5)) at a $[A_2]_0/[A_1]_0$ ratio of 6. $k_2/(k_1[A_1]_0)$ = 100 (*a*), 1,000 (*b*), 10,000 (*c*), 100,000 (*d*). Markers represent values obtained by the steady state formula in Eq. (3.9)

In effect, the conclusion drawn using the concept of the rate determining step was derived here independently, through the use of steady state concentrations. The usual condition of using this line of thought is that the typical time scale of the consumption of the steady state intermediate should be a lot shorter than the time scale of its production. In the example dealt with thus far, this means that $k_2 \gg k_1[A_1]_0$ should hold.

Now returning to the paradoxical nature of the steady state assumption, Eq. (3.8) is better understood in a broader context. In fact, the rate of concentration change for the steady state intermediate is not literally zero, it is just much smaller than the rates of concentration change for the reagents and products:

$$-\frac{d[A_1]}{dt} = -\frac{d[A_2]}{dt} \gg \left|\frac{d[A_3]}{dt}\right| \ll \frac{d[A_4]}{dt} \qquad (3.11)$$

If the concentration of A_3 is small compared to other substances, then its absolute rate of concentration change must also be small compared to others. Conservation equations in this system require the following sum to be zero at any time (this can simply be demonstrated by summing the corresponding lines from the rate equation):

$$\frac{d[A_1]}{dt} + \frac{d[A_3]}{dt} + \frac{d[A_4]}{dt} = 0 \qquad (3.12)$$

The inequality written in Eq. (3.11) means that the middle term can be dropped from Eq. (3.12). Therefore, $k_1[A_1][A_2] - k_2[A_3] \approx 0$ will hold at any time.

In every case, the steady state approach involves selecting a minor intermediate and substituting the differential equation for its concentration change by a direct functional relationship with other concentrations appearing in the system. In the previous example, this led to the same conclusion as using the concept of the rate limiting step. Generally, this equivalence is true when a series of irreversible reactions is considered. The real utility of the steady state approach is manifested in systems containing reversible reactions.

Consider the following scheme, in which the reverse reaction of the first process is added to Eq. (3.5):

$$A_1 + A_2 \underset{k_2}{\overset{k_1}{\rightleftharpoons}} A_3 \overset{k_3}{\longrightarrow} A_4 \tag{3.13}$$

The steady state assumption can be used readily for species A_3 provided that the inequality $k_2 + k_3 \gg k_1[A_1]_0$ holds. The following expression is obtained for the steady state concentration of the intermediate:

$$[A_3]_{ss} = \frac{k_1}{k_2 + k_3}[A_1]_t[A_2]_t \tag{3.14}$$

The rate of product formation and reactant loss is then given as usual, by substituting the (approximate) steady state concentration into the (precise) rate equation:

$$\frac{d[A_4]}{dt} = k_3[A_3]_{ss} = \frac{k_1 k_3}{k_2 + k_3}[A_1][A_2] \tag{3.15}$$

There is no obvious way of deriving this rate equation through the use of rate determining steps. Indeed, k_1 might not determine the rate of product formation any more despite $k_3 \gg k_1[A_1]_0$, as some of the intermediate formed in the first process is diverted from participating in product formation by the reverse reaction.

Equation (3.15) can be easily transformed to a single-concentration rate equation and solved because $[A_2]_t = [A_1]_t - [A_1]_0 + [A_2]_0$ holds for every time instant, and the rate of loss of A_1 is equal to the rate of A_4 formation. In fact, Eq. (3.15) can be rearranged into a form that is identical to the rate equation of the mixed second order process (see Eq. (2.26)), the only difference is the designation of the rate constant (k_1 in Eq. (2.26) and $k_1 k_3/(k_2 + k_3)$ in Eq. (3.15)). The time dependence of the concentration of the product is therefore given as:

$$[A_4] = \frac{[A_2]_0[A_1]_0(1 - e^{-k_1 k_3([A_2]_0-[A_1]_0)t/(k_2+k_3)})}{[A_2]_0 - [A_1]_0 e^{-k_1 k_3([A_2]_0-[A_1]_0)t/(k_2+k_3)}} \tag{3.16}$$

Jim Espenson's book [9] also describes a variant of the steady state approach, which is referred to as the **improved steady state method** [18]. The use of this method will be illustrated on the example given in Eq. (3.13). The line of thought leading to (3.14) is used here as well. However, this equation is differentiated further to give the following formula (the fact that the derivatives of the concentrations of A_1 and A_2 are equal must be remembered here):

$$\frac{d[A_3]_{ss}}{dt} = \frac{k_1}{k_2 + k_3}([A_1] + [A_2])\frac{d[A_1]}{dt} \tag{3.17}$$

The formulas in Eqs. (3.17) and (3.15) are then substituted into Eq. (3.12) to obtain the following improved expression:

$$\frac{d[A_1]}{dt} + \frac{k_1}{k_2 + k_3}([A_1] + [A_2])\frac{d[A_1]}{dt} + \frac{k_1 k_3}{k_2 + k_3}[A_1][A_2] = 0 \tag{3.18}$$

This equation can be easily rearranged into a single concentration rate equation form if the connection between the concentrations of A_1 and A_2 is also taken into account:

$$\frac{d[A_1]}{dt} = -\frac{k_1 k_3 [A_1]([A_1] - [A_1]_0 + [A_2]_0)}{k_2 + k_3 + k_1[A_1] + k_1([A_1] - [A_1]_0 + [A_2]_0)} \tag{3.19}$$

An implicit solution to this particular rate equation can actually be found, so the concentration of product can be estimated as a function of time.

An important use of the steady state approximation occurs in deriving the Michaelis–Menten equation in enzyme kinetics based on the Briggs-Haldane mechanism, which involves the reversible association of two species followed by the formation of the final product, which also reproduces one of the reagents [5, 19]:

$$A_1 + A_2 \underset{k_2}{\overset{k_1}{\rightleftharpoons}} A_3 \overset{k_3}{\longrightarrow} A_1 + A_4 \tag{3.20}$$

The widely used notation for the Briggs-Haldane mechanism is that A_1 is an enzyme (E), A_2 is a substrate (S), A_3 is a substrate-enzyme adduct (ES), whereas A_4 is the final product (P) formed from the substrate. This is a highly instructive example to be handled for demonstrating the use of the steady state approach, so the details of the solution process will be elaborated at some length here.

A steady state assumption for A_3 yields the same equation as Eq. (3.14). However, conservation of mass now requires the following:

$$\frac{d[A_1]}{dt} + \frac{d[A_3]}{dt} = 0 \tag{3.21}$$

Therefore, the concentrations of A_1 (enzyme) and A_3 (substrate-enzyme adduct) are linked directly. Typically, $[A_3]_0 = 0$, so the conservation equation can be used in the following manner:

$$[A_1]_0 = [A_1] + [A_3]_{ss} = [A_1] + \frac{k_1}{k_2 + k_3}[A_1][A_2] \tag{3.22}$$

Expressing the concentration of A_1 from this equation gives:

$$[A_1] = \frac{[A_1]_0}{1 + \frac{k_1}{k_2+k_3}[A_2]} \qquad (3.23)$$

Therefore, the final formula for the concentration change of species A_2 and A_4 is given as:

$$-\frac{d[A_2]}{dt} = \frac{d[A_4]}{dt} = k_3[A_3]_{ss} = \frac{k_3[A_1]_0[A_2]}{\frac{k_2+k_3}{k_1} + [A_2]} \qquad (3.24)$$

This rate equation is now analogous to Eq. (2.12) in Chap. 2, which was called Michaelis–Menten rate equation there ($k_a = k_3[A_1]_0$ and $k_b = (k_2 + k_3)/k_1$). The two characteristic parameters are customarily called maximum rate ($v_{max} = k_3[A_1]_0$) and the Michaelis constant ($K_M = (k_2 + k_3)/k_1$). The parameter v_{max} is the maximum rate to which the rate of product formation converges at high concentrations of the substrate, whereas K_M is the substrate concentration at which the rate of substrate formation is half of the v_{max} value. Although the Briggs–Haldane mechanism features three rate constants (k_1, k_2, and k_3), only two parameters (v_{max} and K_M) can be resolved from the data. There is no way of determining k_1 and k_2 separately.

The success of the steady state approximation is typically a consequence of the limited concentration sensitivity. Yet, even when the minor intermediate can be detected by an unusually sensitive technique, the steady state approximation will still be useful to derive the concentration change of reactants and products, and also give a highly reasonable approximation for the concentration of the steady state intermediate except in a short initial period. Furthermore, the steady state approach necessarily involves loss of information: for each minor species in steady state, the number of resolvable parameters decreases by one. These resolvable parameters are often not rate constants, but their combinations.

3.5 Pre-Equilibrium Approach

Consider the following scheme of an initial reversible first order reaction followed by an irreversible first order reaction.

$$A_1 \underset{k_2}{\overset{k_1}{\rightleftharpoons}} A_2 \xrightarrow{k_3} A_3 \qquad (3.25)$$

This scheme can be derived from Eq. (2.68) by setting $k_4 = 0$ and the solution can also be obtained in this way from Eq. (2.69). Figure 3.5 shows the time dependence of the concentration of A_3 under conditions when $[A_2]_0 = [A_3]_0 = 0$

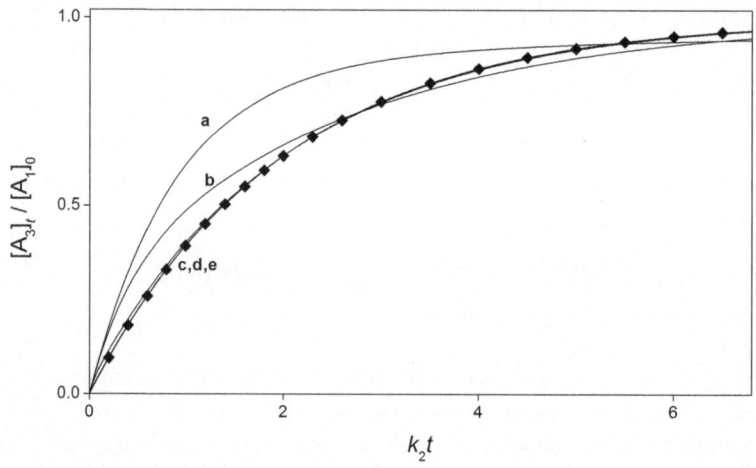

Fig. 3.5 Scaled concentrations as a function scaled time for final product A_3 in the scheme described by Eq. (3.25) at the constant k_1/k_2 ratio of 1. $k_1/k_3 = k_2/k_3 = 0.1$ (*a*), 1 (*b*), 10 (*c*), 100 (*d*), 1,000 (*e*). Markers represent values obtained by the pre-equilibrium formula in Eq. (3.9)

and $k_1/k_2 = 1$. As shown, the traces are independent of the actual value of k_1/k_3 when $k_1/k_3 = k_2/k_3 > 10$.

The curves with constant k_1/k_2 become independent of $k_1/k_3 (= k_2/k_3)$ when the time scales of the two processes are very different. Under conditions like this, the first reaction can be considered to be in equilibrium all the time on the time scale of the second process. Therefore, the concentrations of A_1 and A_2 are connected by the following equation:

$$[A_2] = \frac{k_1}{k_2}[A_1] \tag{3.26}$$

Substituting this into the full rate equation of the process yields the following differential equation:

$$\frac{d[A_1]}{dt} + \frac{d[A_2]}{dt} = \frac{d[A_1]}{dt} + \frac{k_1}{k_2}\frac{d[A_1]}{dt} = -k_3[A_2] \tag{3.27}$$

Rearrangement of the previous equation yields the following formula:

$$\frac{d[A_1]}{dt} = -\frac{k_3 k_1}{k_2 + k_1}[A_1] \tag{3.28}$$

The solution of this rate equation for the product formation is described by a simple exponential function with a composite first order rate constant that can be calculated from the values of k_1, k_2 and k_3:

$$[A_3]_t = [A_1]_0 \left(1 - e^{\frac{k_3 k_1}{k_2 + k_1} t} \right) \tag{3.29}$$

This approach is called the **pre-equilibrium approximation**. It is somewhat similar to the steady state approach, but the validity range is very different. It is not limited to minor intermediates. The success of the pre-equilibrium approach is usually a consequence of the limited time resolution of the experimental data. Technically, it implies that an intermediate is already present at $t = 0$. Obviously, such an approach is not useful to predict the time dependence of the build-up of this intermediate when this process is within the time resolution of the experiments. Furthermore, one cannot hope to resolve the rate constants of both the forward and reverse reactions of a pre-equilibrium. The information content of such experiments is limited to the ratio of these two rate constants, which is the same as the equilibrium constant of the process. To make this point clear, schemes often do not even indicate the two rate constants separately on the arrows, only an equilibrium constant (K) is used. So it is quite customary to write Eq. (3.25) in the following form if the use of the pre-equilibrium approach is emphasized:

$$A_1 \overset{K}{\rightleftharpoons} A_2 \overset{k_3}{\longrightarrow} A_3 \tag{3.30}$$

The improved steady state approach presented in the previous section can be used instead of the pre-equilibrium approach each time. As a matter of fact, the improved steady state approach is also successful whenever the steady state approach can be used. The essence of the pre-equilibrium approach is a separation of time scales between subsequent processes (the earlier must be a reversible reaction), whereas the steady state approach makes use of the minor nature of an intermediate. For the improved steady state approach, it is the time separation between the build-up and consumption of an intermediate (and not subsequent reactions!) that is the important aspect. However, the use of the improved steady state approach is not inevitable: if it can be used, either the steady-state approach or the pre-equilibrium approach (or, occasionally, both) will provide essentially the same interpretation with less tedious mathematical details.

The pre-equilibrium approach is typically used for proton transfer reactions in aqueous solution (such as acid dissociation or protonation), which are understood to be among the fastest possible processes. Often times, species that only differ in their protonation state are not even considered as different species, but their concentration is obtained by multiplying the concentration of a combined species that represents the sum of the different form with a mole fraction. There are in fact very few experimental techniques with which such proton transfer reactions can be monitored.

A very interesting case is when there is a fast equilibrium that involves the formation of some sort of an adduct from one of the initial substances and one of the products of a relatively slow reaction. A simplified scheme can be given as follows:

$$A_1 \xrightarrow{k_1} A_2$$

$$A_1 + A_2 \overset{K}{\rightleftharpoons} A_3 \tag{3.31}$$

The wording pre-equilibrium is not a very fortunate one for this case, as the equilibrium does not precede the reaction of interest. Yet the same principles can be used here as well: the concentration of species A_3 can be given as the function of the concentrations of A_1 and A_2 at each time. This scheme can be a source of greatly counterintuitive phenomena. One of the most extreme kinetic curves ever seen by the author of this book describes the time dependence of the concentration of triiodide (I_3^-) ion in the redox reaction between dithionate ion and periodate ion, which is given as:

$$[I_3^-]_t = 0.5 \times (k_a - k_b(t - k_c)^2) + 0.5 \times |k_a - k_b(t - k_c)^2| \tag{3.32}$$

For those interested in chemical kinetics, there is great joy in discovering why this formula is suitable for describing the kinetic curve despite the fact that it has very little general bearing on kinetics because of the extreme rarity of this case. The reader will not be deprived of an opportunity of experiencing this joy on her or his own after consulting the literature [15].

3.6 Relaxation Kinetics

Relaxation kinetics is a special approach to describing kinetic curves that emerged out of necessity when very fast reactions were studied. The time in which two solutions can be mixed seems to have some very fundamental limitations. Therefore, reactions with typical time scales below 1 ms can only be investigated if they can be initiated without mixing two liquids. The temperature, the pressure of a system, the properties of the electric field can be changed on a microsecond time scale, highly intense lasers make it possible to use energetic light pulses faster than a nanosecond to initiate chemical reactions. With a few exceptions, these extremely fast kinetic methods only work in studying equilibrium reactions and the initiating change only causes a minor disturbance of the equilibrium. The system then converges back to its state of equilibrium, this process is called **relaxation**.

In Sect. 3.1, it was already remarked that at the end of chemical processes, flooding conditions typically prevail even in cases where the initial concentrations are comparable. In relaxation kinetics, all of the concentrations are flooded, not even the limiting reagent is an exception. (Incidentally, the concept of limiting reagent is a very murky one for equilibrium reactions because the process stops before the "limiting reagent" is consumed.) So the changes in the concentrations are minor compared to the values of the concentrations in these cases.

To illustrate this phenomenon, consider the following scheme of the reversible mixed second order reaction as written in Eq. (2.50):

$$A_1 + A_2 \underset{k_2}{\overset{k_1}{\rightleftharpoons}} A_3 + A_4 \tag{3.33}$$

The rate equation of this process is handled in more detail in Chap. 2. It is advantageous to formulate the rate equation using the distance from the equilibrium concentration values, $x_t = [A_1]_t - [A_1]_\infty$. As pointed out earlier, the concentration changes are a lot smaller than the concentrations under relaxation conditions, so $x_t \ll [A_i]_\infty$ holds for any species. The rate equation shown earlier in Eq. (2.53) can be simplified then by the following approximation:

$$\begin{aligned}\frac{dx}{dt} &= -(k_1[A_1]_\infty + k_1[A_2]_\infty + k_2[A_3]_\infty + k_2[A_4]_\infty)x + (k_2 - k_1)x^2 \\ &\approx -(k_1[A_1]_\infty + k_1[A_2]_\infty + k_2[A_3]_\infty + k_2[A_4]_\infty)x\end{aligned} \tag{3.34}$$

Clearly, the time dependence of x is given by an exponential function with a first order rate constant that can be calculated from the equilibrium concentrations and the second order rate constants. The reciprocal of this rate constant is commonly referred to as the **relaxation time** of the process. The experimental study usually involves changing the equilibrium concentrations systematically, determining the relaxation time and formulate the rate equation based on the results.

If more than one equilibrium reaction is involved, the concentration changes are described by combinations of exponential functions. Usually, the number of exponential terms and the number of relaxation times will be equal to the number of different equilibria involved in the scheme. Needless to say, this gives rise to a huge variability of schemes used in relaxation kinetics, where the use of multiexponential functions is almost exclusive. An entire book could, and in fact was, written about relaxation kinetics [3].

3.7 Chain Reactions

Chain reactions are special processes in chemical kinetics. They usually involve multiple reactive intermediates, which are transformed to each other in turns while the reactants are transformed to products. They usually involve a higher degree of complexity of the chemical scheme than the systems dealt with thus far. As an example, consider the following scheme:

$$A_1 \underset{k_2}{\overset{k_1}{\rightleftharpoons}} A_2$$
$$A_2 + A_3 \xrightarrow{k_3} A_4 + A_5 \qquad (3.35)$$
$$A_4 + A_6 \xrightarrow{k_4} A_2 + A_7$$

The overall result of the scheme in Eq. (3.35) is that the reaction between A_3 and A_6 occurs producing products A_5 and A_7 without any direct interaction between A_3 and A_6. Species A_2 and A_4 are typically highly reactive from a chemical point of view and they are minor intermediates in the process. Reaction k_3 produces A_4 from A_2, whereas reaction k_4 produces A_2 from A_4, so the two reactive species participate in a continuous cycle. This type of mechanisms is often referred to as a **chain reaction** in chemical kinetics. It is quite advantageous and visually appealing to depict chain reactions in the form of a cycle as shown in Fig. 3.6.

The kinetic treatment of chain reactions has well-established rules, which are mostly of mathematical rather than experimentally observed nature. First, there are different types of reactions in a chain:

* **Initiation**: This kind of step involves the formation of chain carriers from species that do not participate in the cycle (or at least are not chain carriers themselves). The number of chain carriers increases in this step.
* **Propagation**: A step that transforms one chain carrier to another. No change in the number of chain carriers occurs.
* **Termination**: The consumption of a chain carrier to products which are not chain carriers themselves. The number of chain carriers decreases.

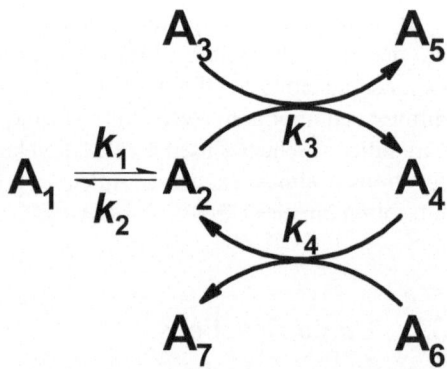

Fig. 3.6 A visual representation of the chain reaction described in Eq. (3.35)

- **Branching**: This step involves chain carriers both as reactants and products, but the number of chain carriers formed is higher the number of chain carriers consumed. This step is distinct from initiation, because at least one chain carrier is a reactant in a branching process.

Initiation, propagation, and termination are necessary parts of all chain reactions. Branching may or may not occur. When it is present, it usually adds immense complication to the mathematical description of a system. Furthermore, reactions not related to the chain may also occur in reaction systems that are otherwise described as chain reactions. It is also possible that a chain reaction features multiple initiation or termination steps.

The mathematical handling of chain reactions typically relies on steady state assumptions for the chain carriers. In the example of Eq. (3.35), steady state equations are set up for A_2 and A_4 as follows:

$$\frac{[A_2]}{dt} = k_1[A_1] - k_2[A_2]_{ss} - k_3[A_2]_{ss}[A_3] + k_4[A_4]_{ss}[A_6] = 0$$

$$\frac{[A_4]}{dt} = k_3[A_2]_{ss}[A_3] - k_4[A_4]_{ss}[A_6] = 0$$

(3.36)

These are linear equations, which can be solved easily. In this case, adding these two equations eliminates the unknown $[A_4]_{ss}$ and makes it possible to calculate $[A_2]_{ss}$ directly, which, in turn, leads to a formula for the steady state concentration of the other chain carrier:

$$[A_2]_{ss} = \frac{k_1}{k_2}[A_1]$$

$$[A_4]_{ss} = \frac{k_1 k_3 [A_1][A_3]}{k_2 k_4 [A_6]}$$

(3.37)

By using these steady state concentrations, it can be shown that the rates of concentration change for species A_3, A_5, A_6, and A_7 are as follows:

$$-\frac{[A_3]}{dt} = \frac{[A_5]}{dt} = -\frac{[A_6]}{dt} = \frac{[A_7]}{dt} = \frac{k_1 k_3}{k_2}[A_1][A_3]$$

(3.38)

The concentration of reactant A_1 does not change overall in the system. Therefore, the entire process involves the reaction of A_3 with A_6 to produce A_5 and A_7, and species A_1 plays the role of the catalyst.

Although the derivation based solely on the steady state assumption presented in the previous paragraphs is quite useful and is fully capable of deriving all useful kinetic expressions in a chain reaction, this is not the way most experienced kineticists typically think about the phenomenon. There are a set of easily remembered rules that can accelerate deriving the steady state rate expressions greatly. These rules are as follows:

- All the propagation steps have the same rates.
- The rate of the initiation step is equal to the rate of the termination step.
- The rate of the consumption of reagents or formation of products is the same as the rate of a propagation step (inclusion of stoichiometric factors may be necessary).

These simple rules follow from the steady state assumptions, but there is a somewhat more intuitive way of demonstrating their validity. Were the rates of propagation steps not equal to each other, one of the chain carriers would quickly accumulate. Were the rates of the initiation and termination reaction steps different, the number of chain carriers would increase beyond any limit or decrease to zero rapidly. Actually, for a very short initial period, initiation is faster than termination, but the typical experimental methods do not have the time resolution and the concentration sensitivity to pick up this part of the kinetic traces. Finally, the rate of a chain propagation step is identical to the rate of the overall reaction, as reactants are only consumed and products are only formed in propagation steps.

A characteristic of the example shown in Eq. (3.35) is that the chain carriers are not formed from the overall reactants, so the reactants (A_3 and A_6) are not involved in the initiation. This is not necessarily true in all chain reactions. If one (or both) of the reactants is involved in the initiation, its rate of concentration change can still be calculated based on propagation steps only because the rate (not the rate constant!) of the propagation steps is typically much larger than the rate of initiation. This element of the treatment of chain reactions is called the **long chain assumption**. The **average chain length** (n_{chain}) can be calculated as the rate of any propagation step divided by the rate of the initiation step. In the example of Eq. (3.35), the formula is:

$$n_{chain} = \frac{k_3}{k_2}[A_3] \tag{3.39}$$

The efficiency of a chain reaction can be measured by the average chain length, which is basically the average number of cycles a chain carrier participates in before being diverted into a termination step.

Typically, the different reaction steps in a chain reaction cannot be studied independently of each other, and this causes a certain degree of complexity. A very special case is when the initiation reaction is a photochemical process. The overall system in this case is called a **photoinitiated chain reaction**. Such a process gives an opportunity to do some independent studies of the subsystem not involving the initiation by following the changes immediately after switching off the light (which is equivalent to switching off the initiation). An example of the use of this technique is found in the literature [12].

At first sight, it seems that each chain reaction should contain at least two propagation steps. It may be surprising, but in exceptional cases, a single propagation step is already enough to build a chain mechanism. The formal example is:

$$A_1 \underset{k_2}{\overset{k_1}{\rightleftharpoons}} A_2$$

$$A_2 + A_3 \overset{k_3}{\longrightarrow} A_2 + A_4$$

(3.40)

In this case, chain carrier A_2 occurs on both the reactant and product side of reaction k_3. It may seem that this actually leaves A_2 unchanged, but there may be molecular symmetry reasons for this lack of overall change. An experimental example of such a chain mechanism with a single chain carrier is known [13].

Finally, it must not be left without notice that the phrase chain reaction is also in use in other fields of science with a regrettably different meaning. For example, in nuclear fission, chain reaction is often understood to mean that more neutrons are produced in the fission process than necessary to initiate it. In the terms of chemical kinetics, neutrons would be chain carriers in this process, but the essential property is a sign of branching. Branching typically leads to the final outcome that the chain carriers attain high concentrations and will not be minor intermediates any more. Also, because chain carriers are highly reactive, this would imply extremely high reactions rates–an explosion.

3.8 Clock Reactions

In usual chemical reactions, the initial rates of concentration changes are the largest because the concentrations of the reactants are largest at $t = 0$. However, there are cases when this is not true. For kineticists, it is often fascinating to see examples where a reactant is consumed at a rate that accelerates for some time. Conversely, acceleration of the formation rate of a product is also considered to be interesting. This can be taken to the extreme when a product is not formed observably for a relatively long time, then suddenly appears, which may seem quite dramatic if the product is colored. An experimental example of such a process, called the Landolt reaction after the German discoverer, has been known for more than a century [14].

To offer a simple interpretation of the case of the suddenly appearing product, consider the following scheme, in which a first order process is followed by the overall second order reaction with another reagent:

$$A_1 \overset{k_1}{\longrightarrow} A_2$$

$$A_2 + A_3 \overset{k_2}{\longrightarrow} A_4$$

(3.41)

Although trying to solve the rate equation of this scheme analytically might not be an entirely hopeless venture (the time dependence of A_1 is given by an

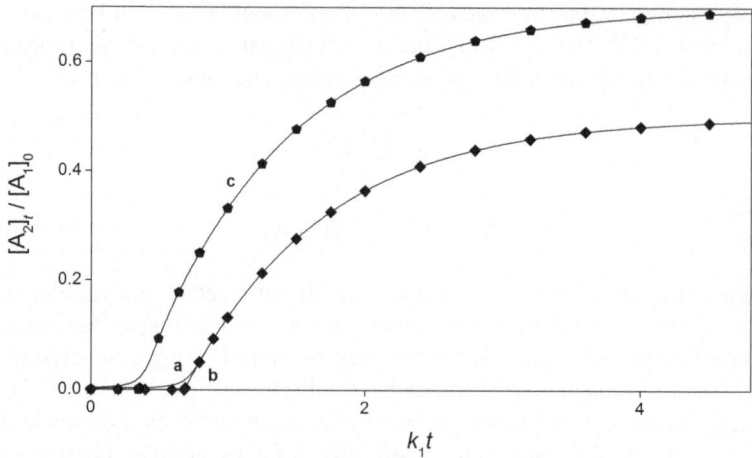

Fig. 3.7 Scaled concentrations as a function scaled time for species A_2 in the scheme described by Eq. (3.41). $[A_3]_0/[A_1]_0 = 0.5$ (*a, b*), 0.3 (*c*). $k_2[A_3]_0/k_1 = 500$ (*a*), 5,000 (*ab*), 300 (*c*). Markers represent values obtained by the pre-equilibrium formula in Eq. (3.43)

exponential function), this book will present some curves that were calculated by a numerical method. In this scheme, A_1 and A_3 are reactants, whereas A_2 and A_4 are products. Therefore, the typical initial conditions include the absence of all products with $[A_2]_0 = 0$. Numerically simulated scaled kinetic curves for the appearance of A_2 for three different case can be seen in Fig. 3.7.

Very little A_2 is formed for a considerable time. The concentrations in this initial region are not actually 0, but they are so small that would be difficult to pick up by any experimental method. This time period is called the **clock time** or **Landolt time**. When the initial concentration of A_3 is larger ($[A_3]_0/[A_1]_0 = 0.5$), the clock time is longer and the final concentration of A_2 reached is lower. A lower $[A_3]_0$ ($[A_3]_0/[A_1]_0 = 0.3$ in the example of Fig. 3.7) leads to a shorter clock time and larger final concentration of A_2. The calculated curve shapes do not depend on the ratio $k_2[A_3]_0/k_1$, provided that this ratio is high enough.

The qualitative interpretation is clear. For the rate constant of the two steps, $k_2[A_3]_0 \gg k_1$ holds in all cases. So basically step k_1 is rate determining for the formation of A_4 until reactant A_3 is present. Species A_2 is a minor intermediate in this time interval. When practically all A_3 is consumed, A_2 becomes a final product and appears suddenly.

As the concentration of A_1 is easy to derive analytically, the time necessary for the total consumption of A_3, which is identical to the clock time, is estimated quite easily from the equation $[A_3]_0 = [A_1]_0 - [A_1]_{t_{clock}}$. This clock time is given as follows:

$$t_{clock} = \frac{1}{k}\ln\left(\frac{[A_1]_0}{[A_1]_0 - [A_3]_0}\right) \qquad (3.42)$$

The time dependence of the concentration of A_2 is estimated based on the previous lines of thought:

$$[A_2]_t = 0 \quad \text{if } t \leq t_{\text{clock}}$$

$$[A_2]_t = ([A_1]_0 - [A_3]_0)(1 - e^{-k_1(t - t_{\text{clock}})}) \quad \text{if } t \geq t_{\text{clock}}$$

(3.43)

The markers in Fig. 3.7 show how Eq. (3.43) approximates the actual kinetic curve. The agreement is quite good. As stated previously, the only condition for this approximation to work well is $k_2[A_3]_0 \gg k_1$.

The clock time interval introduced in this section is also commonly termed as an **induction period** or **incubation period**. The interpretation shown here is primarily dependent on a stoichiometric condition ($[A_3]_0 < [A_1]_0$) and a kinetic condition (the ratio of rate constants). These are the two typical criteria for designating a process of a **clock reaction**. However, induction periods in the formation of the product may arise for other reasons as well [16]:

- Autocatalytic processes typically have a part where the rate of product formation accelerates (see curve -0.985 in Fig. 2.3).
- In a consecutive reaction mechanism, the product(s) of the second or later steps appear with some sort of delay only (see curve 0.2 in the inset of Fig. 2.5).
- A branching chain reaction may allow the concentrations of reactive intermediates to build up to non-minor levels and cause highly accelerating product formation rate (see the end of Sect. 3.7).
- In a thermal explosion, the uncontrolled temperature increase as a consequence of heat formation in the process has a role similar to that of an autocatalyst.

The concepts of clock reaction and **clocklike behavior** were analyzed in a rigorous manner in a recent publication [11].

3.9 Competition Kinetics and Chemical Clocks

Consider the following scheme already discussed at some length earlier as Eq. (2.108) in Chap. 2:

$$A_1 + A_3 \xrightarrow{k_1} A_4$$

$$A_2 + A_3 \xrightarrow{k_2} A_5$$

(3.44)

This scheme is often described as a **competition** of reagents A_1 and A_2 for the common species A_3. Although the concentrations of A_1 and A_2 could not be given as a function of time in Chap. 2, a useful formula connecting $[A_1]$ and $[A_2]$ to each other was still derived in Eq. (2.111). It is typically possible to measure the concentrations of products A_4 and A_5 in this example. So Eq. (2.111) can be rearranged to show $[A_4]$ and $[A_5]$ instead of the remaining reactant concentrations:

$$\frac{k_2}{k_1} = \frac{\ln\left(1 - \frac{[A_5]_t}{[A_2]_0}\right)}{\ln\left(1 - \frac{[A_4]_t}{[A_1]_0}\right)} \tag{3.45}$$

This is a very instructive equation. First of all, it is valid for any time. So determining the concentration of A_4 and A_5 at a single time instant is enough to estimate the ratio of the two rate constants. Of course, the reaction should not be allowed to go as far as to consume both A_1 and A_2 completely. So if one of the two rate constants is known from an independent source, the value of the other can be calculated without measuring any time dependence. This is why the method explained here is called a **chemical clock**. Note that this phenomenon has nothing to do with clock reactions, which were interpreted in the previous section.

The concentration of reactant A_3, although it is the goal of the competition between reagents A_1 and A_2, does not even appear in Eq. (3.45). Therefore, A_3 may even be formed as the reaction progresses. The mathematical description does not change if A_3 is consumed in reactions not listed in the scheme. So A_3 can even be a minor intermediate that never builds up to detectable concentrations. Such highly reactive species are often radicals (i.e., chemical species with an odd overall number of electrons). Therefore, the method is also called **radical clock**.

With the concept of the chemical clock, it is sufficient to determine the rate constant of just one of the reactions of a reactive intermediate (in an exceptionally fortunate case, for example), and then it will be possible to determine a whole series of further rate constants. If the rate constant of the reference reaction is very different from the (unknown) rate constant of the target reaction, changing the initial concentrations leave a lot of room for the experimenter. A further simplification of the formula in Eq. (3.45) is possible if the processes are only followed to low conversion so that the approximation $\ln(1 - [A_5]_t/[A_2]_0) \approx -[A_5]_t/[A_2]_0$ is valid. Under these conditions, Eq. (3.45) is simplified into the following form:

$$\frac{k_2}{k_1} \approx \frac{[A_5]_t}{[A_4]_t} \tag{3.46}$$

Several different methods could be devised based on the principles introduced in this section: measuring the concentrations ratios of substances can in fact give a lot of information on rate constants relative to some reference data.

3.10 Parameter Insensitivity and Indistinguishable Schemes

The question of **parameter sensitivity** has already arisen in the previous sections of this chapter. With the concept of the rate determining step, steady state approach, or pre-equilibrium approach, some of the rate constants appearing in a kinetic scheme quite naturally cannot be determined. This is a more general phenomenon and is primarily caused by the fact that the theoretical prediction of the detected data does not depend on the value of some of the parameters. These parameters are called insensitive. The phenomenon may be observed even when no approximations are used, but the evaluation is based on the numerical integration of the entire assumed kinetic scheme.

Insensitivity occurs in three different forms. The first is when an upper limit can be given for the value of the parameter as a conclusion. Any values lower than this limit will be suitable to describe the experimental data. If the parameter is a rate constants of an elementary step, such an elementary step can usually be simply omitted from the scheme without any effect on the quality of the fit. The second case is when a lower limit is established for the parameter as a conclusion. Any value higher than the limit will interpret the data equally well. The most common reason for this phenomenon is that a reaction step is not within the time resolution of the experimental method. The third case is when neither of two parameters can be determined, but a certain combination of them (such as the ratio or the product) is clearly determined by the data. The reasons for such insensitivity may be quite diverse.

Indistinguishable schemes are different sets reaction steps that interpret the same set of data equally well (the phrase **equivalent kinetic expressions** is also used in this context). They typically arise from the fact that not every concentration in a system can be determined to any reasonable precision. Consider the following consecutive scheme where reactants A_1 and A_2 are involved in a fast pre-equilibrium to form a reactive intermediate A_3, which reacts with a third reagent A_4 to give a product in a subsequent step:

$$A_1 + A_2 \overset{K}{\rightleftharpoons} A_3$$
$$A_3 + A_4 \overset{k_2}{\longrightarrow} A_5 \tag{3.47}$$

If A_3 is a minor intermediate, the formation rate of product A_5 can be approximated as follows:

$$\frac{d[A_5]}{dt} = k_2 K [A_1][A_2][A_3] \tag{3.48}$$

Now consider a different scheme, where A_1 and A_4 are involved in the pre-equilibrium, and the final product is formed in the reaction of minor intermediate A_3 and reactant A_2:

$$A_1 + A_4 \overset{K}{\rightleftharpoons} A_3$$
$$A_3 + A_2 \overset{k_2}{\longrightarrow} A_5 \tag{3.49}$$

Again, if A_3 is a minor intermediate, the rate of product formation can be given by Eq. (3.48). Therefore, the two presented mechanisms cannot be distinguished by following the concentration of the final product A_5. The distinction is also impossible if reactants A_1, A_2, and A_4 are selectively followed, because the intermediate is minor.

In aqueous solutions, a very typical occurrence of indistinguishable schemes is called **proton ambiguity**. As remarked earlier, proton transfer reactions in water are among the fastest known processes. It is well known from introductory chemistry that protons only appear in chemical systems attached to some other species. This protonation reaction is typically a fast pre-equilibrium. Proton ambiguity means that solely based on kinetic data, it is impossible to tell which of the reagents the proton binds to in this pre-equilibrium.

Occasionally, it is possible to make an informed decision between equivalent schemes based on the feasibility of the parameter values that are determined in them. For example, second order rate constants have upper limits because of diffusion (see Sect. 4.1 in Chap. 4). If a scheme would predict that a second order rate constant is larger than the diffusion limited value, the scheme itself can be rejected with high certainty. Other such limitations may also serve as tests of the validity of indistinguishable schemes.

Sometimes it is possible to resolve questions arising from parameter insensitivity or indistinguishable schemes by better experiment design. A systematic theoretical study of the proposed scheme will usually reveal what additional tests must be done. However, experimental observations are typically limited by factors more fundamental than the ingenuity of the researcher.

3.11 Statistical Kinetics

Statistical kinetics describes cases when some sort of coincidence of parameter values disguises kinetic phenomena in a counterintuitive way. Unlike the methods described in the previous sections of this chapter, this is not a question of approximations, but the exact solutions of the scheme conspire to mislead the investigator because of a mathematical coincidence.

Examples of this phenomenon will be shown using the scheme of two consecutive irreversible processes already discussed in Chap. 2 as Eq. (2.57):

$$A_1 \xrightarrow{k_1} A_2 \xrightarrow{k_2} A_3 \qquad\qquad (3.50)$$

The solution of this scheme (Eq. (2.59)) is generally a first order decay for the concentration of A_1 and a double exponential function for $[A_2]$ and $[A_3]$. Even in Chap. 2, some attention was called to the fact that the coincidence $k_1 = k_2$ leads to a different form of the exact solution, which was given as a first order polynomial multiplied by an exponential function in Eq. (2.60).

A similar case occurs in the scheme of a mixed second order reaction followed by a first order process (Eq. (2.87)), in which the coincidence $k_2 = k_1([A_1]_0 - [A_2]_0)$ gave rise to an entirely different formula for the solution (Eq. (2.90)).

However, statistical kinetics involves some more important and also more counterintuitive examples. To understand the origins of these, the scheme of two consecutive irreversible processes is considered again but now for the more general case of $k_1 \neq k_2$. If an instrumental signal is used to follow the reaction, it might very well happen that all of the species participating in the reaction contribute to the detected property. As pointed out in Chap. 1, absorption measurements are very common in chemical kinetics. So the absorbance change at a given wavelength (λ) can be described in this system as follows (cf. Eq. (1.21)):

$$A_{t,\lambda} = \varepsilon_{1,\lambda}[A_1]_t + \varepsilon_{2,\lambda}[A_2]_t + \varepsilon_{3,\lambda}[A_3]_t \qquad (3.51)$$

Combining this equation with the exact solution for the concentration changes given in Eq. (2.60) yields:

$$A_{t,\lambda} = \frac{\varepsilon_{1,\lambda}(k_1 - k_2) - \varepsilon_{2,\lambda}k_1 + \varepsilon_{3,\lambda}k_2}{k_1 - k_2}[A_1]_0 e^{-k_1 t} + (\varepsilon_{2,\lambda} - \varepsilon_{3,\lambda})\left(\frac{[A_1]_0 k_1}{k_1 - k_2} + [A_2]_0\right) e^{-k_2 t}$$
$$+ ([A_1]_0 + [A_2]_0 + [A_3]_0)\varepsilon_{3,\lambda}$$
$$(3.52)$$

Although this formula describes a double exponential function, certain combinations of parameters will make it impossible to detect one of the two exponential terms and, therefore, will make it look like a single exponential curve is detected. An obvious such coincidence is $\varepsilon_{2,\lambda} = \varepsilon_{3,\lambda}$, which gives zero as the multiplying factor before the exponential term k_2. In this case, the second process does not change the signal at all, which explains why k_2 cannot be determined.

A more nuanced coincidence is $(\varepsilon_{1,\lambda} - \varepsilon_{2,\lambda})k_1 = (\varepsilon_{1,\lambda} - \varepsilon_{3,\lambda})k_2$, which makes the k_1 term undetectable. This relationship between molar absorptivities and rate constants might seem quite rare, but it is inevitable that this should be valid at certain isolated wavelengths. In such cases, changing the wavelength will reveal the double exponential nature of the entire process. However, it is actually not uncommon that this coincidence occurs simultaneously at all wavelength values in cases when multiple equivalent reaction centers and absorbing moieties are involved in a reaction. A high number of experimental examples are known (some of them more complicated than the one shown here) in the literature [1, 2, 20].

Finally, it should be noted that the phrase statistical kinetics is occasionally used in a very different sense in the scientific literature, to refer to the particle based approach to chemical kinetics, which was called stochastic kinetics in the first chapter (see Sect. 1.1), and contrasted with the concentration based deterministic kinetics that is the subject of this book.

3.12   A Qualitative Approach to Chemical Kinetics

Although describing concentrations as a function of time is the primary tool in
chemical kinetics, other information about the nature of concentration changes in
a system may often be valuable. There is **qualitative approach** to studying the
structure of various models, which focuses on the directions of changes and the
tendencies. This short section only aims to summarize the problems for which this
sort of approach may be helpful. The interested reader is referred to Chap. 4 of the
textbook of Érdi and Tóth [8] for further information.

The qualitative approach does not give explicit formulas for temporal changes,
but in fact the results are often quantitative. A similar phenomenon was already
encountered in Chap. 2, where the schemes in Eqs. (2.104) and (2.108) could not be
solved analytically, but useful formulas relating the concentrations to each other
were derived. In this sense, the present section is somewhat misplaced in this
Chapter because the methods used do not necessarily involve any approximations.

Stationary, oscillatory, and chaotic behaviors have central roles in the qualitative
approach. These phenomena cannot typically be observed in closed systems, the
usual reactor used is a CSTR (see Sect. 1.4). As the studied processes were
counterintuitive for classical chemical kinetics, they are also referred to as **exotic
kinetic phenomena**.

Stationary conditions are said to prevail if no concentration change occurs in a
system. The concentrations at which this is possible is a **stationary point**, which is
a solution of the following equation:

$$f_i([A_1], [A_2], \dots, [A_n]) = 0 \tag{3.53}$$

Obvious examples of stationary conditions in closed systems are when one
of the reagents is totally consumed or when the state of chemical equilibrium
is reached in a reversible process. These rarely produce any exotic phenomena.
In a CSTR, however, there are far more interesting stationary states, which are
classified based on their stability. Intuitively, the stability of a stationary point
depends on whether the system returns to this state after a small perturbation or
not. A common technique to answer these questions is called **Lyapunov stability
analysis**. A system may show several stationary points, this is a phenomenon called
multistationarity. In such cases, it is typically an interesting question how the
system may proceed from one stationary state to another.

In certain cases, some concentrations in a system may be periodic functions
of time ($[A_i]_{t+T} = [A_i]_t$ for any t). This behavior is called **oscillation**. The
qualitative theory of reaction kinetics offers a reasonably good understanding about
the conditions necessary for such behavior to emerge.

It is also possible that some concentrations in a system are neither periodic
nor tend to a stationary point. Loosely speaking, this behavior is called **chaotic**.
Although such systems are not unknown in chemistry, they typically have much
higher significance in other fields of science concerned with temporal changes.

References

1. Algarra, A.G., Fernández-Trujillo, M.J., Basallote, M.G.: A DFT and TD-DFT Approach to the understanding of statistical kinetics in substitution reactions of M3Q4 (M=Mo, W; Q=S, Se) cuboidal clusters. Chem. Eur. J. **18**, 5036–5046 (2012)
2. Armstrong, F.A., Henderson, R.A., Sykes, A.G.: Kinetic studies on reactions of iron-sulfur proteins. 3. Oxidation of the reduced form of clostridium pasteurianum 8-iron ferredoxin with inorganic complexes. Observation of single-stage kinetics for a difunctional protein reactant. J. Am. Chem. Soc. **102**, 6545–6551 (1973)
3. Bernasconi, C.F.: Relaxation Kinetics. Academic Press, New York (1976)
4. Bodenstein, M.: Die Zersetzung des jodwasserstoffgases im licht. Z. Phys. Chem. **22**, 23–33 (1897)
5. Briggs, G.E., Haldane, J.B.: A note on the kinetics of enzyme Action. Biochem. J. **19**, 338–339 (1925)
6. Dunn, B.C., Meagher, N.E., Rorabacher, D.B.: Resolution of stopped-flow kinetic data for second-order reactions with rate constants up to 10^8 $M^{-1}s^{-1}$ involving large concentration gradients. experimental comparison using three independent approaches. J. Phys. Chem. **100**, 16925–16933 (1996)
7. Edelson, D.: The lew look in chemical kinetics. J. Chem. Educ. **52**, 642–644 (1975)
8. Érdi, P., Tóth J.: Mathematical Models of Chemical Reactions. Theory and Applications of Deterministic and Stochastic Models. Manchester University Press, Manchester (1989)
9. Espenson, J.H.: Chemical Kinetics and Reaction Mechanisms, 2nd edn. McGraw-Hill, New York (1995)
10. Farrow, L.A., Edelson, D.: The steady-state approximation: fact or fiction? Int. J. Chem. Kinet. **6**, 787–800 (1974)
11. Horváth, A.K.; Nagypál, I.: Classification of clock reactions. Chem. Phys. Chem (2007), DOI: 10.1002/cphc.201402806
12. Kerezsi, I., Lente, G., Fábián, I.:Highly efficient photoinitiation in the cerium(iii)-catalyzed aqueous autoxidation of sulfur(IV). An example of comprehensive evaluation of photoinduced chain reacions. J. Am. Chem. Soc. **127**, 4785–4793 (2005)
13. Kingston, J.V., Verkade, J.G., Espenson, J.H.: 3/2-Order chain kinetics involving a postulated dicationic intermediate in the isomerization of a p-isocyano to a p-cyano azaphosphatrane monocation. J. Am. Chem. Soc. **127**, 15006–15007 (2005)
14. Landolt, H.: Ueber die Zeitdauer der Reaction zwischen Jodsäure und schwefliger Säure. Ber. Dtsch. Chem. Ges. **19**, 1317–1365 (1886)
15. Lente, G., Fábián, I.: Effect of dissolved oxygen on the oxidation of dithionate ion. Extremely unusual kinetic traces. Inorg. Chem. **43**, 4019–4025 (2004)
16. Lente, G., Bazsa, G., Fábián, I.: What is and What isn't a clock reaction? New J. Chem. **31**, 1707–1707 (2007)
17. Lin, C.T., Rorabahcer, D.B.: Mathematical approach for stopped-flow kinetics of fast second-order reactions involving inhomogeneity in the reaction cell. J. Phys. Chem. **78**, 305–308 (1974)
18. McDaniel, D.H., Charles R., Smoot, C.R.: Approximations in the kinetics of consecutive reactions. J. Phys. Chem. **60**, 966–969 (1956)
19. Michaelis, L., Menten, M.L.: Die kinetik der invertinwirkung. Biochem. Z. **49**, 333–369 (1913)
20. Vanderheiden, D.B., King, E.L.: Solvolysis of iodopentaaquachromium(III) ion in acidic aqueous dimethyl sulfoxide. J. Am. Chem. Soc. **95**, 3860–3866 (1973)

Chapter 4
Information from Parameter Values

In general, the values of kinetic parameters, mostly rate constants, should be determined in the course of investigations, but they do not have high significance compared to the rate equation. The relation between rates and concentrations takes precedence over the numerical values of parameters in establishing mechanisms or finding the elementary reactions in a system. Yet every now and then interpreting the values themselves is a source of information. There are some absolute limitations on these values, or they can be compared with other parameters obtained in independent measurements to test certain assumptions. These tests are quite powerful in falsifying certain ideas: if a rate constant is outside the reasonable range, or an equilibrium constant obtained from kinetic results is significantly different from that measured in direct equilibrium studies, the researcher should look for an erroneous assumption in her or his line of arguments.

Some theoretical interpretation of reactivity or reaction rates is also based on the values of parameters (typically rate constants). Here, a comparison of measured values with theoretical predictions is an important technique in validating theory.[1] Today, the theory of activation is usually used to interpret reactivity, which will be discussed in some detail in this chapter.

4.1 Diffusion Controlled Processes

The rates of chemical reactions involving the interaction of two different particles must have some sort of highest possible value because the speed at which the particles can approach each other is physically limited. This line of thought is quite general and can be used to establish an upper limit for the rate constants of

---

[1]This author believes that an experiment cannot be wrong, but its information content can be seriously misinterpreted.

© Gábor Lente 2015
G. Lente, *Deterministic Kinetics in Chemistry and Systems Biology*,
SpringerBriefs in Molecular Science, DOI 10.1007/978-3-319-15482-4_4

bimolecular reactions. As the process in which particles approach each other is called diffusion in both the gas and condensed phases, this theoretical maximum is called the **diffusion controlled rate constant**. Note the fact that the limit is actually set for a bimolecular reaction (which is an elementary reaction), but not for second order processes in general.

For gas phase reactions, this issue is easily dealt with based on **collision numbers,** which are readily available from the kinetic-molecular theory of gases. For a mixture of two different particles (A_1 and A_2), the number of collisions in unit time and unit volume in a container is given by the following equation:

$$Z_{A_{1,2}} = N_A^2 \sigma_{A_{1,2}} \sqrt{\frac{8 k_B T}{\pi \mu_{A_{1,2}}}} [A_1][A_2] \tag{4.1}$$

In this formula, N_A is the Avogadro constant, σ_{A_1,A_2} is the average cross section of particles A_1 and A_2, k_B is the Boltzmann constant, whereas μ_{A_1,A_2} is the reduced mass of particles A_1 and A_2, which is defined as follows:

$$\mu_{A_{1,2}} = \frac{m_{A_1} m_{A_2}}{m_{A_1} + m_{A_2}} \tag{4.2}$$

The notations m_{A_1} and m_{A_2} mean the respective masses of particles A_1 and A_2. A comparison of a second order rate law with Eq. (4.1), taking into account the fact that concentration is measured in moles rather than particle numbers, yields the diffusion controlled bimolecular gas phase rate constant as follows:

$$k_{\text{gasdiff}} = N_A \sigma_{A_{1,2}} \sqrt{\frac{8 k_B T}{\pi \mu_{A_{1,2}}}} \tag{4.3}$$

A typical value for the diffusion limited rate constant could be estimated at room temperature ($T = 298$ K) by selecting particles A_1 and A_2 to be spheres with a radius of 100 pm ($\sigma_{A_{1,2}} = \pi (10^{-7} \text{m})^2 = 3.1 \times 10^{-14}$ m^2) and a molar mass of 50 g/mol ($\mu_{A_{1,2}} = m_{A_1}/2 = 0.05$ kg/mol$/6.0 \times 10^{23}$ mol$^{-1}/2 = 4.2 \times 10^{-26}$ kg). In this case, the value $k_{\text{gasdiff}} = 9.3 \times 10^{12}$ mol$/(\text{m}^3\text{s}) \approx 10^{10}$ M^{-1}s^{-1} is obtained, which is also a good general estimate under normal conditions.

Surprisingly, some quite reliably measured bimolecular rate constants in the gas phase are actually higher than the upper limit calculated from Eq. (4.3) and the physical dimensions of the particles. These are interpreted by assuming that the reactants do not satisfy one of the important assumptions of the kinetic-molecular theory of gases, namely the absence of attractive interactions between particles. If particles attract each other, they may react with each other even if their paths do not collide, but only approach each other. In cases like this, it is common to use Eq. (4.3) in a reverse way, which means that ($\sigma_{A_{1,2}}$) is actually calculated from the measured value of the second order rate constant rather than the physical dimensions of the particles. This value is called the **reactive cross section** in the process, and can be thought of as the size of the surface of one of the particles that serves as a target for the other one. Reactive cross sections actually carry the same information as a second order rate constant. An analogous quantity is in widespread use to characterize the efficiency of nuclear reactions.

Estimating the diffusion controlled second order rate constant is more of a challenge for solution reactions. A commonly used [22, 25], albeit not completely satisfactory line of thought will be presented here based on the **Fick equation**, which states that the flux (J) of a chemical is directly proportional to the gradient in its concentration (i.e., the spatial derivate perpendicular to the surface through which the flux is measured). In the following equation, this is formulated for reactant A_1:

$$J_{A_1} = -D_{A_1} \frac{d[A_1]}{dx} \tag{4.4}$$

The proportionality constant in the Fick equation, D, is called the diffusion constant. Consider the space around a single particle of the other reactant A_2. There is no reason to assume that any directions have special roles, so the concentration of A_1, denoted $[A_1]_r$ here, can be described as a function of r, the distance measured from particle A_2 (i.e., the system shows spherical symmetry). When this distance is the sum of the radii of the two particles ($r_{A_{1,2}}$), reaction occurs, so the concentration $[A_1]_{r_{A_{1,2}}}$ is zero. On the other hand, at a large distance from the selected A_2 particle, the concentration of A_1 is the same as its bulk concentration $[A_1]_\infty = [A_1]$.

The total flow (Φ) of reactant A_1 toward a single particle of A_2 can be obtained by multiplying the flux with the surface area, which is the surface of the sphere with radius r now:

$$\Phi = D_{A_1} 4\pi r^2 \frac{d[A_1]}{dr} \tag{4.5}$$

This is a separable different equation for dependent variable $[A_1]_r$ with independent variable r (although not an autonomous one). The solution, first considering $[A_1]_\infty = [A_1]$ as an initial condition, is easily obtained:

$$[A_1]_r = [A_1] - \frac{\Phi}{4\pi D_{A_1} r} \tag{4.6}$$

However, there is another condition set by the physical assumptions: $[A_1]_{r_{A_{1,2}}} = 0$. Substituting this condition into Eq. (4.6) yields an equation from which a unique value for the total flow Φ is originated:

$$\Phi = 4\pi D_{A_1} r_{A_{1,2}} [A_1] \tag{4.7}$$

The number of independent A_2 particles in the unit volume of the reactor is $N_A[A_2]$. However, to calculate the overall rate of concentration change, another remark must be made. Thus far, particles of A_2 were considered to be stationary.

In fact, they also diffuse in the system. A common solution to this problem, which is only slightly better than ignoring it altogether, is to multiply Φ by two to obtain the macroscopic rate. It should be noted that a fully analogous problem is also encountered when collision numbers are calculated in the kinetic-molecular theory of gases. This is the reason why the reduced mass of the two particles is used in Eq. (4.1) instead of the actual masses. Bearing all previous sequences of thought in mind, the rate of concentration change for species A_1 is as follows:

$$-\frac{d[A_1]}{dt} = N_A[A_2]\Phi = 8\pi D_{A_1} r_{A_{1,2}} N_A[A_1][A_2] \qquad (4.8)$$

Therefore, the diffusion limited second order rate constant in solution is obtained as:

$$k_{\text{soldiff}} = 8\pi D_{A_1} r_{A_{1,2}} N_A \qquad (4.9)$$

Equation (4.9) could be used by substituting typical values of diffusion constants and molecular sizes to obtain a reasonable estimate of a diffusion controlled rate constant in solution. However, experience shows that diffusion constants are typically only dependent on the size of the diffusing particles and the viscosity of the medium they move in. The **Stokes–Einstein equation** connects the diffusion constant, the size of the particle and the viscosity of the solvent (η) as follows [9, 10, 23]:

$$D_{A_1} = \frac{k_B T}{6\pi r_{A_1} \eta} \qquad (4.10)$$

If the two particles are assumed to have approximately the same radius (so that $r_{A_{1,2}} = 2r_{A_1}$) and the gas constant R is used instead of the product of the Avogadro constant (N_A) and Boltzmann constant (k_B), the expression for the diffusion limited rate constant is greatly simplified into the following formula:

$$k_{\text{soldiff}} = \frac{8RT}{3\eta} \qquad (4.11)$$

The viscosity of water is about 9×10^{-4} Pa s at room temperature, so the diffusion controlled limiting rate constant is estimated to be 7×10^6 m^{3}mol^{-1}s^{-1} under these conditions, which is 7×10^9 M^{-1}s^{-1} if the concentration unit of M is used. This shows a remarkable, and most probably purely accidental, agreement with k_{gasdiff}. On a somewhat offhand note, it should be pointed out the temperature dependence of this value typically comes from the temperature dependence of the viscosity (a factor of 6 for water from its freezing point to its boiling point) rather than from the explicit appearance of T in the formula (a factor of 1.4 in the same range).

Now some thoughts should be devoted to understanding what aspects of this derivation are unsatisfactory. First of all, it uses a spatially continuous notion of concentration to interpret a phenomenon (finite collision frequency of species) that is clearly a manifestation of the particulate nature of matter. This is not simply a systematic error, this is a conceptual error. Furthermore, the derivation predicts that the concentration of one reagent will be depleted in the surroundings of the other, which would mean selective inhomogeneity in the solution. Equation (4.6) implies that at a distance of $10r_{A_{1,2}}$ from a particle of A_2, the concentration of A_1 is still only 90 % of the bulk value. An $r_{A_{1,2}}$ value of 300 pm is not uncommon, this would mean spheres of depletion with radii 3 nm in typical cases. Even at a low concentration of 1 mM, these spheres of depletion would take up some 7 % of the entire volume. From a still not very high concentration of 15 mM, the entire volume is in the these spheres of reactant depletion. Finally, it is totally unclear how the proximity of reactant A_1 could influence the local concentration of A_2 at distances which are too high for a reaction to take place.

Given these problems, the derived value of $k_{soldiff}$ shown in Eq. (4.11) must involve an uncertainty of at least an order of magnitude. If a bimolecular rate constant of 10^{12} $M^{-1}s^{-1}$ is required in a series of elementary steps, it is safe to say that the diffusion limit falsifies this value and an alternative interpretation must be thought. However, values of 4×10^{10} $M^{-1}s^{-1}$ or 6×10^{10} $M^{-1}s^{-1}$ cannot be ruled out in this way.

As a final remark in this section, it should be emphasized that diffusion can only limit the rate constants of bimolecular elementary reactions. Unimolecular (and therefore first order) processes are not subject to the same limitation. However, it is more than probable that some other physical limits, such as the typical time scale of molecular vibrations, determine a fastest possible unimolecular rate constant as well. For example, the Eyring equation presented later in Eq. (4.14) would give $k_B T / h = 6.2 \times 10^{12}$ s^{-1} for a barrierless ($\Delta^{\ddagger} G = 0$) reaction at 298 K, which could be thought of as a limit for unimolecular processes.

4.2   Activation and Transition States

In the previous section, efforts were reported to calculate the rate at which different particles approach each other in solution or gas phases. However, it is an experimental fact that most determined second order rate constants are much lower than $k_{gasdiff}$ or $k_{soldiff}$. The conclusion from this fact is clear: only a tiny fraction of the encounters of the two reactants in an elementary reaction actually lead to change.

Another interesting observation is that most determined rate constants increase with temperature in a fashion that is close to exponential, whereas the same dependence for the diffusion limited rate constants is at most linear, but often even less sensitive. This leads to the conclusion that a greatly higher fraction of the encounters leads to chemical change at higher temperatures.

A commonly used equation for describing the temperature dependence of rate constants was published by Svante Arrhenius [1], the third winner of the Nobel prize in chemistry. Today, it is called the **Arrhenius equation** and is typically formulated as follows:

$$k = Ae^{-\frac{E_a}{RT}}$$ (4.12)

In Eq. (4.12), k is a rate constant, E_a is typically called **activation energy**, A is called **pre-exponential factor** or **frequency factor**, R is the gas constant, T is the (absolute) temperature. One should note that A has the same units as the rate constant, so the name frequency factor is only reasonable for first order rate constants with the dimension of inverse time.

Current thinking about the Arrhenius equation [6, 17, 18] is that the reason why it became so commonly used is not the fact that it fitted to the temperature-dependent rate constant data best among a set of functions tried for this purpose. The popularity owes much to a very appealing interpretation of the activation energy as the extra energy needed in collisions that lead to a chemical reaction. This concept is easy to give in an attractive visual way, which is called **reaction energy profile**. An example is shown in Fig. 4.1.

Unfortunately, the highly visual nature of this energy profile is at the expense of the scientific information content. The x axis is often called "reaction coordinate", but typically lacks any meaningful definition. The type of energy shown on the y axis is seldom specified further, although it would thermodynamically important to distinguish between internal energy, enthalpy, free energy, or perhaps other energy-related quantities. So both axes in Fig. 4.1 leave room for (well-deserved) scientific criticism. In addition, the only energy values that have physical meaning in this graph is the **reactant state** (beginning), activation energy (maximum), and **product state** (end). It is not uncommon to indicate only these three energy values in the

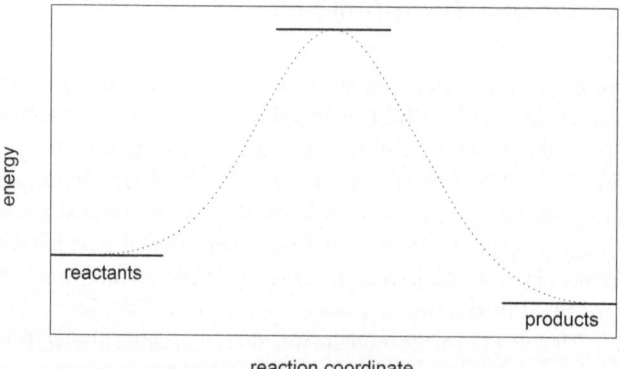

Fig. 4.1 An example of a reaction energy profile

way the solid lines do in Fig. 4.1. Yet, it is even more customary to draw the highly arbitrary curve connecting these three states, as given by the dotted line of Fig. 4.1.

The activation energy is represented in Fig. 4.1 as the difference between the energy of the reactant state and the energy maximum. The overall energy change of the reaction is the difference between the product and reactant states.

It is not impossible to define the axes in an energy profile in a much more satisfactory manner. Electron energy (sometimes, after certain corrections, even free energy) can be calculated for all possible spatial arrangements of nuclei in a reactive system. The values of this energy as a function of nuclear coordinates is called the **potential energy surface** (it is usually a hypersurface, meaning that it typically has more than two independent variables). The reactant state and the product state are both well-defined minima on this surface. The lowest lying maximum of all possible curves connecting the reactant and product states, which is named a **saddle point** in mathematical topology, is called the **transition state**. The arrangement of nuclei in this state is sometimes referred to as the activated complex. So the E_a activation energy appearing in the Arrhenius equation can be related to the energy difference between the transition and the reactant state.

This is the point where the concept of **activation** was born. Activation is the process in which the reactant state gains the extra energy needed to reach the transition state. It is well known that energy is generally not uniformly distributed between particles, but it is possible that a fraction of particles have higher energies than the average. In statistical thermodynamics, the Boltzmann distribution is often used to predict what fraction of particles has an energy higher than a pre-set value. The essence of the **Boltzmann distribution** is an exponential term with reciprocal temperature, $e^{-E/RT}$, so the Arrhenius equation actually relies on this energy distribution, which is thought to be quite general in nature.

Once the concept of the transition state is introduced, other physical properties of this state can also be estimated. These are called **activation parameters** and they are often thought to carry some diagnostic information about the molecular details of a reaction. It is the opinion of this author that activation parameters are seriously overvalued in today's chemical kinetics. Often times, information content is attributed to them even when it is not present at all. Activation only makes sense for an elementary reaction. However, activation parameters for nonelementary reactions are often discussed and conclusions are drawn, which, needless to say, lacks any reasonable theoretical support. Activation is a scientific theory that has a validity range, researchers should remind themselves of this fact every now and then. For example, there are compelling arguments against the use of the concept of activation in solid state reactions [19].

In addition to the Arrhenius equation, which was primarily based on experimental observations, there are other formulas in widespread use to interpret the temperature dependence of rate constants, and some of them have elaborate theoretical backgrounds. In the gas phase, **collision theory** is often used. This simply posits that a second order rate constant can be estimated by multiplying the rate constant characterizing the number of collisions (i.e., the diffusion controlled rate constant,

Eq. (4.3)) by the probability term from the Boltzmann distribution, which gives the fraction of successful collisions. Eventually, the following formula is obtained:

$$k = k_{\text{gasdiff}} e^{-\frac{E_a}{RT}} = N_A \sigma_{A_{1,2}} \sqrt{\frac{8k_B T}{\pi \mu_{A_{1,2}}}} e^{-\frac{E_a}{RT}} \qquad (4.13)$$

Note that this equation can only be used to describe the rate constant of a bimolecular process. It can be considered as a two-parameter function, in which reactive cross section $\sigma_{A_{1,2}}$ (see the explanation after Eq. (4.3)) and activation energy E_a may be determined from experimental data. The reactive cross section can be compared with estimates on the actual geometrical cross section of the particles.

In modern solution phase kinetics, the **Eyring equation** [12–14] is the most popular way of interpreting the temperature dependence of a rate constant and calculating activation parameters. This formula has detailed theoretical background, which is deeply routed in quantum mechanics and is often called transition state theory. The equation can be given in the following form:

$$k = \frac{k_B T}{h} e^{-\frac{\Delta^{\ddagger} G}{RT}} = \frac{k_B T}{h} e^{-\frac{\Delta^{\ddagger} H}{RT} + \frac{\Delta^{\ddagger} S}{R}} \qquad (4.14)$$

In Eq. (4.14), k_B is the Boltzmann constant as in previous equations, whereas h is the Planck constant, $\Delta^{\ddagger} G$ is the **standard free energy of activation**, $\Delta^{\ddagger} H$ is the **standard enthalpy of activation**, $\Delta^{\ddagger} S$ is the **standard entropy of activation**.[2] A property called **transmission factor**, denoted κ, is often given as an additional multiplying term in the Eyring equation, yet its value is seldom obtained in any meaningful theoretical way. Experimentally, it is probably best viewed as a contributing factor to the entropy of activation, from which it cannot be separated.

It should be emphasized that dimensional analysis of Eq. (4.14) shows that it yields a first order rate constant, so the interpretation is done only for a unimolecular reaction. In practice, it is also very common to use the Eyring equation for bimolecular rate constants, but separate lines of thought are necessary to validate this process. In a bimolecular process, the two reactants are thought to form an adduct ($A_1 \cdot A_2$) first, whose concentration is calculated using the pre-equilibrium approach (see Sect. 3.5). Then the process itself is interpreted as the unimolecular reaction of this adduct, which is present at a very low concentration. In this way, the second order rate constant can be obtained by multiplying the equilibrium constant for adduct formation ($K^{\ddagger}$) with the first order rate constant characterizing the unimolecular reaction of the adduct. As the Eyring equation is primarily about

---

[2]The notations $\Delta^{\ddagger} G$, $\Delta^{\ddagger} H$, and $\Delta^{\ddagger} S$ follow IUPAC recommendations. In actual use, the forms $\Delta G^{\ddagger}$, $\Delta H^{\ddagger}$, and $\Delta S^{\ddagger}$ are a lot more common in the literature.

giving the temperature dependence of a rate constant, the temperature dependence of the equilibrium constant K must be described at this step. There is a well-known formula for this in thermodynamics, which gives the basis of the van't Hoff equation. There is some unfortunate mismatch between the usual conventions of thermodynamics and kinetics because the former usually requires dimensionless equilibrium constants expressed by activities, whereas the latter prefers using concentrations and equilibrium constants with physical dimensions. This mismatch will be solved here by including a standard concentration ($c^{\ominus}$) in the formula giving the temperature dependence of the equilibrium constant:

$$K^{\ddagger} = \frac{[A_1 \cdot A_2]}{[A_1][A_2]} = \frac{1}{c^{\ominus}} e^{-\frac{\Delta G^{\ominus}}{RT}} = \frac{1}{c^{\ominus}} e^{-\frac{\Delta H^{\ominus}}{RT} + \frac{\Delta S^{\ominus}}{R}} \qquad (4.15)$$

In this equation, $\Delta G^{\ominus}$, $\Delta H^{\ominus}$, and $\Delta S^{\ominus}$ are the **standard free energy**, **enthalpy**, and **entropy** changes of the adduct formation, in order. Combining Eqs. (4.15) and (4.14) gives an interpretation of a bimolecular rate constant based on the Eyring equation:

$$k_{\text{bim}} = K^{\ddagger}k = \frac{k_B T}{c^{\ominus} h} e^{-\frac{\Delta^{\ddagger} H + \Delta H^{\ominus}}{RT} + \frac{\Delta^{\ddagger} S + \Delta S^{\ominus}}{R}} \qquad (4.16)$$

Typically, the sums $\Delta^{\ddagger} H + \Delta H^{\ominus}$ and $\Delta^{\ddagger} S + \Delta S^{\ominus}$ are not separated but used as the single activation enthalpy and activation entropy for the bimolecular process. Yet, when interpreting their information content, the effect of the pre-equilibrium must not be forgotten. For example, adduct formation reactions typically have substantially negative $\Delta S^{\ominus}$ values, which will be reflected in the determined activation entropy. Furthermore, Eq. (4.16) also emphasizes the fact that for any meaningful use of the Eyring equation for bimolecular processes, the unit of rate constant must be $M^{-1}s^{-1}$ because the usual convention is $c^{\ominus} = 1$ M.

Equations (4.12), (4.13), and (4.14) share the feature that they use at most two parameters to describe the temperature dependence of a rate constant. When rate constants are measured over a wide range of temperatures, using two parameters is often insufficient to obtain a reasonable fit. Therefore, based on purely experimental observations, a three-parameter version of Eq. (4.12), called the **modified Arrhenius equation**, is sometimes employed:

$$k = AT^n e^{-\frac{E_a}{RT}} \qquad (4.17)$$

This equation has the same parameters as the Arrhenius equation in Eq. (4.12), and an additional one denoted n, which is a dimensionless power, and, quite interestingly, lacks a commonly used name. Equation (4.17) is a three parameter function that, as far as the functional form of temperature dependence goes, unites the previously introduced three equations. Setting $n = 0$ is identical to the Arrhenius

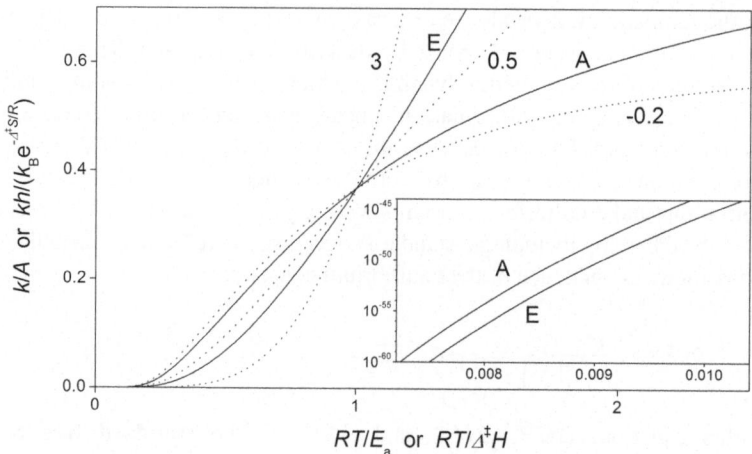

Fig. 4.2 Scaled plots of the rate constant as a function of temperature as described by the Eyring (E, Eq. (4.14)), Arrhenius (A, Eq. (4.12)) and modified Arrhenius equations (3, 0.5, −0.2, Eq. (4.17)). The values 3, 0.5, and −0.2 denote the value of n in the modified Arrhenius equation. *Inset*: the early part of the plot using logarithmic scaling on the y axis

equation in Eq. (4.12), $n = 0.5$ is analogous to the formula obtained in the collision theory (Eq. (4.13)), whereas $n = 1$ corresponds to the Eyring equation in Eq. (4.14) (although without the well-established theoretical background). Figure 4.2 presents a scaled graph that shows the usual dependencies described in these equations.

The plots shown in Fig. 4.2 may be very unusual even for experienced kineticists. The main reason for this weirdness is that the values of untransformed temperature and rate constant are used on the axes. In addition, the scaling is done in a way that the graph shows temperature values at which RT is comparable to E_a or $\Delta^{\ddagger}H$. These would normally be very high temperatures, conditions under which an often overlooked property of the Arrhenius equation manifests: at these temperatures, the rate constant becomes close to independent of the temperature. This phenomenon is easily understood conceptually: under these conditions, all the collisions involve enough energy for a reaction to occur [6]. In practical cases, $RT \ll E_a$ or $RT \ll \Delta^{\ddagger}H$ is almost always true. Therefore, the inset in Fig. 4.2 shows the initial region of the large graph. Note that the rate constant is given on a logarithmic scale here, so it actually increases quite rapidly as temperature increases.

Yet, the inset of Fig. 4.2 is still not the conventional visual way of showing temperature dependencies of rate constants. The activation energy is typically determined by linearizing Eq. (4.12) to give the following rearranged formula:

$$\ln k = \ln A - \frac{E_a}{RT} \tag{4.18}$$

In a linearization method, $\ln k$ is plotted as a function of $1/T$, and E_a is calculated from the (negative) slope of the fitted straight line. This linearized graph is called an

Arrhenius plot. A very similar linearization, unsurprisingly called the **Eyring plot**, is also in widespread use for the Eyring equation:

$$\ln\frac{k}{T} = \ln\frac{k_{\mathrm{B}}}{h} + \frac{\Delta^{\ddagger}S}{R} - \frac{\Delta^{\ddagger}H}{RT} \tag{4.19}$$

Two points should be made here. First, in a strict mathematical sense, a logarithm can only interpreted on dimensionless physical properties. This is less of a problem for Eq. (4.19), where plotting $\ln((k k_{\mathrm{B}})/(hT))$ on the y axis as a function of $1/T$ would give an acceptable solution with only a minor effort. For Eq. (4.18), a formal solution is to plot $\ln(k/k_{\mathrm{ref}})$ as a function of $1/T$ and obtain $\ln(A/k_{\mathrm{ref}})$ as the intercept instead of $\ln A$, where k_{ref} is the value of the rate constant at a suitably chosen reference temperature. In practice, however, using Eqs. (4.18) and (4.19) is typical, and some care is exercised when determining the physical units of the obtained parameters.

The second point is that Eqs. (4.18) and (4.19) are linearized. This is typically not a desirable practice from a statistical point of view, as will be emphasized in Sect. 5.12 of Chap. 5. Yet, this is one of the highly exceptional cases when linearizing does not really have unfavorable effects. Rate constants are typically very sensitive to temperature, so it is not uncommon that the value of k changes more than an order of magnitude over the studied temperature range. Typically, the relative uncertainties of these rate constants are independent of temperature. So if the original, nonlinearized form is used for fitting, it is very important to use proportional weighting (see Sect. 2.3 in Chap. 2). Using the logarithm of the rate constants in fitting without weighting (which is the same as uniform weighting) is almost fully equivalent to using proportional weighting on the untransformed data. Furthermore, the typical temperature range of experimental studies is small compared to the absolute temperature, which means that the statistically undesirable effects of inverting the temperature axis are minimal and can be safely neglected in comparison with other, experimental sources of uncertainty.

Calculating the activation enthalpy and activation entropy from the temperature dependence of rate constants based on the Eyring equation is such a common problem in chemical kinetics that some of the numerical details of fitting will be given in the next paragraphs. This analysis will be based on the linearized form (Eq. (4.19)) and needs no weighting. If rate constants $k_1, k_2, \ldots, k_N$ are measured at temperatures $T_1, T_2, \ldots, T_N$, the least squares fitting yields the enthalpy and entropy of activation as follows:

$$\Delta^{\ddagger}H = R\frac{\sum_{i=1}^{N}\left(\frac{1}{T_i} - \frac{1}{N}\sum_{i=1}^{N}\frac{1}{T_i}\right)\left(\ln\frac{k_i}{T_i} - \frac{1}{N}\sum_{i=1}^{N}\ln\frac{k_i}{T_i}\right)}{\sum_{i=1}^{N}\left(\frac{1}{T_i} - \frac{1}{N}\sum_{i=1}^{N}\frac{1}{T_i}\right)^2} \tag{4.20}$$

$$\Delta^{\ddagger}S = \frac{R}{N}\sum_{i=1}^{N}\ln\frac{k_i}{T_i} - R\ln\frac{k_{\mathrm{B}}}{h} - \frac{\Delta^{\ddagger}H}{N}\sum_{i=1}^{N}\frac{1}{T_i}$$

During the course of data analysis, the standard deviations of these quantities must always be calculated and an appropriate number of significant figures must be reported. The standard deviations can be calculated by the following formulas:

$$\sigma_{\Delta^\ddagger H} = R \sqrt{\frac{\sum_{i=1}^{N}\left(\ln\frac{k_i}{T_i} - \frac{\Delta^\ddagger S}{R} - \ln\frac{k_B}{T} + \frac{\Delta^\ddagger H}{RT}\right)^2}{\frac{1}{N}\sum_{i=1}^{N}\left(\frac{1}{T_i} - \frac{1}{N}\sum_{i=1}^{N}\frac{1}{T_i}\right)^2}}$$

$$\sigma_{\Delta^\ddagger S} = R \sqrt{\sum_{i=1}^{N}\left(\ln\frac{k_i}{T_i} - \frac{\Delta^\ddagger S}{R} - \ln\frac{k_B}{T} + \frac{\Delta^\ddagger H}{RT}\right)^2} \sqrt{\frac{1}{N} + \frac{\left(\sum_{i=1}^{N}\frac{1}{T_i}\right)^2}{N\sum_{i=1}^{N}\left(\frac{1}{T_i} - \frac{1}{N}\sum_{i=1}^{N}\frac{1}{T_i}\right)^2}}$$

$$(4.21)$$

Activation parameters can be defined independently of the theoretical equations presented thus far and are sometimes handled as exclusively experimental properties. A rearrangement of the Arrhenius equation serves as a basis to calculate an activation energy from the temperature derivative of rate constants as follows (k_{ref} is, again, the rate constant at a single preselected reference temperature):

$$E_a = -R\frac{\partial \ln(k/k_{ref})}{\partial(1/T)} \qquad (4.22)$$

This formula emphasizes the fact that the activation energy actually does not carry direct information on the value of a rate constant (see Sect. 5.8): it is the sensitivity to changes in temperature that is of primary importance. With the definition of Eq. (4.22), the activation energy is allowed to show temperature dependence, which actually takes the concept very far away from its theoretical origins.

Similarly, the activation enthalpy can be defined as follows:

$$\Delta^\ddagger H = -R\frac{\partial \ln\left((kT_{ref})/(Tk_{ref})\right)}{\partial(1/T)} \qquad (4.23)$$

If values of this experimental activation enthalpy are found to be temperature dependent, by analogy with thermodynamics, the heat capacity of activation can also be introduced:

$$\Delta^\ddagger C_p = -\frac{\partial \Delta^\ddagger H}{\partial T} \qquad (4.24)$$

Although experimentally determined values of $\Delta^\ddagger C_p$ are occasionally reported in the literature of chemical kinetics, it is highly debatable whether they carry any meaningful information. Furthermore, the high uncertainties involved in numerical derivation have already been mentioned in Sect. 1.5. The activation heat capacity is a second derivative, and it is seldom possible to determine its value within reasonable experimental uncertainty limits.

Activation volume, determined from the pressure dependence of rate constants, is in widespread use in modern research on chemical mechanisms. In the gas phase, partial pressure and concentration are proportional to each other. Therefore, rate constants of elementary processes cannot show a dependence on the pressure very much like gas phase equilibrium constants cannot depend on pressure. In the solution phase, however, an analogy with thermodynamic quantities is used again to define the activation volume as follows:

$$\Delta^{\ddagger}V = -RT\frac{\partial \ln(k/k_{\mathrm{ref}})}{\partial p} \tag{4.25}$$

In a typical setup, the rate constants of an elementary reaction are measured as a function of pressure at constant temperature, and the $\Delta^{\ddagger}V$ is obtained from the slope of the straight line fitted in an $\ln(k/k_{\mathrm{ref}})$ vs. p plot.

The depth of the theoretical interpretation of the activation volume is by no means similar to those of the activation enthalpy and entropy. Quantum chemistry usually works with potential energy surfaces. Therefore, activation enthalpies or even activation entropies have natural counterparts in theory to which they can be compared. As remarked earlier, the Eyring equation has a well-established background in quantum mechanical calculations. Volume is a much less available property in such calculations.

4.3 Electrostatic Effects

Electrostatic forces are usually viewed as strong in chemistry. Even in chemical kinetics, extra considerations are needed to account for electrostatic interactions, although these are usually seen as secondary modifying factors. The starting point of these considerations is typically the **Coulomb law**, which is stated for the force acting between two ions, A_1 and A_2 here:

$$F = \frac{z_{A_1}z_{A_2}Q_e^2}{4\pi\varepsilon r^2} \tag{4.26}$$

In this formula, z_{A_1} and z_{A_2} are the charge numbers on the interacting ions A_1 and A_2, Q_e is the charge of an electron,[3] ε is the permittivity of the medium (if it is vacuum, then the permittivity of vacuum, ε_0 is used), and r is the distance between the two ions.

---

[3]It is much more common to use e to denote the charge of an electron. Yet this text uses e^t very often for the exponential function, so it seemed necessary to avoid possible confusions by selecting the notation Q_e instead of the usual one.

In the gas phase, the energy requirements of ion formation (which means, in essence, a separation of charges from an electrostatic point of view) are prohibitively high. As a consequence, ions do not usually form in the gas phase and even the reactive intermediates are neutral particles, often with an odd number of electrons (radicals).

In solution, however, interactions with other particles facilitate the formation of ions. It is especially so in polar solvents, which are characterized by high ε values in comparison with ε_0. Electrolytic dissociation (formation of ions and consequent charge separations) is therefore quite common in solvents, where not only intermediates, but reactants and products may also be charged. It is a very basic tenet of chemistry that these species are not fundamentally different from non-charged species, and it is sufficient to account for their electrostatic interactions based on the Coulomb law only.

In chemical kinetics, two different cases must be considered when two charged species react with each other (if at least one of the particles is neutral, no electrostatic forces arise). The first case is when reactivity is directly influenced by the Coulomb interaction between the reacting species. The second case is when other, nonreactive ions present in the same solution influence a reaction through modifying the ε value in the Coulomb law.

In the first case, the derivation of diffusion controlled rate constants is affected because electrostatic attraction or repulsion modifies the speed at which different particles approach each other. A first approach to solving this problem adds an electrostatic multiplying factor to Eq. (4.11), which takes the following form then:

$$k_{\text{soldiff}} = f \frac{8RT}{3\eta} \tag{4.27}$$

The factor f is calculated from the following formula, in which the use of the Coulomb law is very easily recognized:

$$f = \frac{\delta}{e^{\delta} - 1} \qquad \delta = \frac{z_{A_1} z_{A_2} Q_e^2}{4\pi \varepsilon k_B T r_{A_{1,2}}} \tag{4.28}$$

The value of f, as calculated for distance $r_{A_{1,2}} = 300$ pm and with the ε of water, is 0.25 for two univalent ions of like charges, 2.6 for two univalent ions of opposite charges, whereas the relevant values for two divalent ions are 8.3×10^{-4} and 9.3.

A similar, but more demanding derivation was reported for a case when A_1 and A_2 do not have electric charges but possess **dipole moments** μ_{A_1} and μ_{A_2} [16][4]:

---

[4]The reader should not confuse the dipole moment used here with the reduced mass defined in Eq. (4.2) despite the fact that, following well-established conventions, the Greek letter μ is used to denote both of these different physical quantities.

$$\ln f = \frac{N_A}{4\pi\varepsilon_0(r_{A_{1,2}})^3}\left(\mu^2_{A_1\cdot A_1} - \mu^2_{A_1} - \mu^2_{A_1}\right)\frac{\varepsilon - \varepsilon_0}{2\varepsilon + \varepsilon_0} \tag{4.29}$$

The notation $\mu_{A_1\cdot A_1}$ means the dipole moment of the transition state here. The appearance of this quantity makes Eq. (4.29) close to worthless. As $\mu_{A_1\cdot A_1}$ is obviously unavailable in a direct experimental way, the only possible use of Eq. (4.29) would be to measure the rate constants (which need not be necessarily diffusion controlled) at different ε values and determine $\mu_{A_1\cdot A_1}$ from this series of experiments. Yet a change in ε typically requires a change of solvent, which has a lot more fundamental effects on a reaction than just a change in the permittivity.

The argument put forward in the previous paragraph is quite general in nature. Attempts to interpret changes in the rate constants of a reaction in different solvents are not uncommon in the literature, and they often use the permittivities of the solvents in some way. However, the idea that it is only the value of ε that influences a rate constant in such a case seems an overly simplistic assumption, and is unacceptable even as a crude approximation. It is almost certain that other, more specific influences originating from the solvent are more significant.

These considerations already lead to the second case listed at the beginning of this chapter, i.e., when the influence of nonreactive species must be considered on a rate constant because of their effect on ε. The solvent water is of specific interest here, as it as an excellent solvent of ionic salts. The Debye–Hückel theory [7] is generally used in thermodynamics to account for the electrostatic effect of unreactive charged species on reactions between ions. An important assumption of this theory is that the unreactive ions do not have any specific effects, but their influence comes solely from their overall charge, which can be characterized by the physical property of **ionic strength** (I):

$$I = \frac{1}{2}\sum_{i=1}^{n} z^2_{A_i}[A_i] \tag{4.30}$$

Ionic strength typically influences the rate constants of a reaction, this fact is often referred to as the kinetic salt effect. It is therefore important to keep the ionic strength of the investigated solutions constant as the rate equation is determined experimentally. This matter is not as trivial as it may seem at first sight, as the reactants also contribute to the ionic strength, so changing the reactant concentrations, which is required to determine the rate equation, automatically changes the ionic strength as well. The usual solution is to flood the system with an inert salt in order to keep the ionic strength practically constant. The dependence of a rate constant on ionic strength is usually described by the **Brønsted–Bjerrum equation** [3, 4], which can be given in the following form:

$$k_I = \gamma k_0 \quad \ln\gamma = \frac{2z_{A_1}z_{A_2}\sqrt{I}\,Q_e^3\sqrt{2\pi N_A}}{(1+\sqrt{I})(4\pi\varepsilon_{\text{solvent}}k_B T)^{3/2}} \tag{4.31}$$

In this equation, k_I is the value of the rate constant at ionic strength I, k_0 means the (extrapolated) rate constants at ionic strength 0, whereas $\varepsilon_{\text{solvent}}$ is the permittivity of the pure solvent. Other quantities appearing in the formula have been defined earlier in this chapter.

A common use of Eq. (4.31) is to measure a rate constant at different values of ionic strength, then plot $\ln k$ as a function of $\sqrt{I}/(1 + \sqrt{I})$. This plot should give a straight line, the slope of which will enable the estimation of $z_{A_1} z_{A_2}$.

4.4 Isotope Effects

Isotopes are atomic nuclei with identical atomic numbers but different mass numbers. So isotopes of the same element differ in the number of neutrons in the nucleus, which does not affect the electron structure, where chemical changes occur. Therefore, the existence of isotopes rarely necessitates any additional considerations in chemistry. In addition, the isotopic abundances of elements are close to constant in nature, which is a further reason why this question does not deserve much attention from kineticists—unless the isotopic abundances are intentionally changed between two experiments in the hope of deducing some useful chemical information.

As already remarked, the electronic structure does not change as a consequence of replacing one isotope with another. This provides an essential advantage in isotopic labeling. The kinetics of isotope exchange was already presented in the scheme of Eq. (2.38). Yet the mass difference between the isotopes can occasionally be significant. For example, the formula derived for the diffusion controlled gas phase rate constant in Eq. (4.3) contains the reduced mass of two particles, which is somewhat dependent on the isotopic constitution. More significantly, the frequency of molecular vibrations is also dependent on the mass of the nuclei, which in turn influences reaction rates if the vibration plays a major role in the process. So, selective use of isotope labeling is in principle suitable to determine whether a certain nucleus is directly involved in the molecular changes. If this is the case, the ratio of the rate constants of the original and isotope-substituted particles, which is called the **kinetic isotope effect** (kie), will be quite different from unity. If this is the case, the kie is called a **primary kinetic isotope effect**. Otherwise, the ratio of the two measured rate constants will be close to one, which is called a **secondary kinetic isotope effect**.

The largest difference in the mass ratio of two stable isotopes is 2 for the isotopes of the element hydrogen. So the highest kinetic isotope effects can be expected as a result of selective protium–deuterium substitution. Such a selective substitution is usually chemically viable and its costs are not prohibitively high, either. In practice, the protium–deuterium substitution is the only one that is commonly used to determine kinetic isotope effects. A primary kinetic isotope effect is usually in excess of 5 for this case, sometimes even higher than 100 if the process involves quantum mechanical **tunneling**. Secondary kinetic isotope effects usually remain lower than 2. Every now and then, the deuterium substituted reactant actually reacts faster than the original. This highly counterintuitive case is called an **inverse kinetic isotope effect**.

Isotopes of other elements are very seldom used in measuring kinetic isotope effects primarily because the immense costs of selective labeling are not worth the minor changes that are expected to be measurable as a result.

4.5 Structure–Reactivity Relationships

One of the holy grails of research in chemical kinetics is to understand how changes in the chemical structure influence reactivity, which is typically measured by a value of a rate constant in a certain reaction. This is next to impossible in an absolute sense, but the chances are better if rate constant changes within a closely related family of reactions are investigated.

Generally, the thermodynamic properties of a reaction are unrelated to the kinetic constants. Yet, it is often possible to find correlations between the rate constants and the standard free energies of enthalpies for a limited selection of processes. These correlations usually use the logarithm of the rate constant, which carries the same information as the activation free energy in the Eyring equation (Eq. (4.14)). The correlations are often linear, hence they are called **linear free energy relations**. Sometimes, it is possible to correlate two different groups of reactions with this technique, if the nature of structural changes is identical. Furthermore, the **Bell–Evans–Polanyi principle** [2, 5, 11] establishes that the activation of energy is linearly related to the standard enthalpy of reaction in a number of distinct groups of elementary reactions.

Probably the most general free energy relation (which is nonlinear) is provided by the **Marcus theory** [20, 21] of electron transfer reactions. This theory derives the rate constant of a single electron transfer reaction between reactants A_1 and A_2:

$$A_1 + A_2 \xrightarrow{k_{A_{1,2}}} A_1^+ + A_2^- \tag{4.32}$$

The charges indicated on the species are meant to represent changes primarily and not absolute charges (i.e., the two reactants are not necessarily neutral). The theory estimates the values of $k_{A_{1,2}}$ as follows:

$$\ln \frac{k_{A_{1,2}}}{\sqrt{k_{A_1} k_{A_2}}} = \frac{-\Delta G^\ominus}{2RT} + \frac{(\Delta G^\ominus / RT)^2}{8 \ln(k_{A_1} k_{A_2} / k_{\text{diff}}^2)} \tag{4.33}$$

In Eq. (4.33), $\Delta G^\ominus$ is the standard free energy of the process, k_{diff} is a diffusion controlled rate constant (e.g., from Eq. (4.9)). The quantities k_{A_1} and k_{A_2} are the electron exchange rate constants for the A_1^+/A_1 and A_2/A_2^- redox couples, as indicated in the following equations:

$$A_1 + A_1^{+*} \xrightarrow{k_{A_1}} A_1^+ + A_1^*$$

$$A_2 + A_2^{-*} \xrightarrow{k_{A_2}} A_2^- + A_2^*$$

(4.34)

The **Hammett correlation** is a very successfully used linear free energy relation in organic chemistry [15]. Its validity is limited for series of compounds containing substituted aromatic rings, on which the nonreactive substituents can be changed systematically. Each substituent is assigned a parameter called the **Hammett substituent constant** (denoted σ, hence the alternative name **Hammett sigma** for the constants) based on the acid dissociation constant of the corresponding substituted benzoic acid derivative. The substituent hydrogen serves as the reference point with $\sigma = 0$. The rate constants of a series of reactions involving differently substituted aromatic compounds are determined and then correlation is sought between the logarithm of the rate constant (for historic reasons, ten-based logarithm is used) and the substituent constants. The graph prepared in this way is called a **Hammett plot**. More often than not, there is a linear correlation between the two quantities in this graph, the slope of which is denoted ρ and called the **reaction constant** (or **Hammett rho**).

A fully analogous method was also developed for some subgroups of organic aliphatic compounds and is now called the **Taft equation** [26, 27]. The relevant **Taft substituent constants** are denoted σ^*, whereas the slope of a successful correlation is ρ^*. An advantage of this method is that it can accommodate steric effects (the physical size of the group of atoms) in addition to changes in the electron structure. The **Swain–Scott** [24] and **Edwards** [8] equations are more restricted in scope, but still have some general utility among the linear free energy relations used for organic reactions.

References

1. Arrhenius, H.: Zur Theorie der chemischen Reaktionsgeschwindigkeit. Z. Phys. Chem. **28**, 317–335 (1899)
2. Bligaard, T., Norskov, J.K., Dahl, S., Matthiesen, J., Christensen, C.H., Sehested, J.: The Brønsted–Evans–Polanyi relation and the volcano curve in heterogeneous catalysis. J. Catal. **224**, 206–217 (2004)
3. Bjerrum, N.: Theory of chemical reaction velocity. Z. Phys. Chem. **108**, 82–100 (1924)
4. Brønsted, J.N., Teeter Jr., C.E.: Kinetic salt effect. J. Phys. Chem. **28**, 579–587 (1924)
5. Brønsted, J.N.: Acid and basic catalysis. Chem. Rev. **5**, 231–338 (1928)
6. Carroll, H.F.: Why the Arrhenius equation is always in the "Exponentially Increasing" region in chemical kinetic studies. J. Chem. Educ. **75**, 1186–1187 (1998)

7. Debye, P., Hückel, E.: Zur Theorie der Elektrolyte. I. Gefrierpunktserniedrigung und verwandte Erscheinungen. Phys. Z. **24**, 185–206 (1923)
8. Edwards, J.O.: Polarizability, basicity and nucleophilic character. J. Am. Chem. Soc. **78**, 1819–1820 (1956)
9. Edward, J.T.: Molecular volumes and the Stokes–Einstein equation. J. Chem. Educ. **47**, 261–270 (1970)
10. Einstein, A.: Über die von der molekularkinetischen Theorie der Wärme geforderte Bewegung von in ruhenden Flüssigkeiten suspendierten Teilchen. Ann. Physik **17**, 549–560 (1905)
11. Evans, M.G., Polanyi, M.: Inertia and driving force of chemical reactions. Trans. Faraday Soc. **34**, 11–24 (1938)
12. Eyring, H.: The activated complex in chemical reactions. J. Chem. Phys. **3**, 107–114 (1935)
13. Eyring, H.: The activated complex and the absolute rate of chemical reactions. Chem. Rev. **17**, 65–77 (1935)
14. Eyring, H.: The calculation of activation energies. Trans. Faraday Soc. **34**, 3–11 (1938)
15. Hammett, L.P.: Some relations between reaction rates and equilibrium constants. Chem. Rev. **17**, 125–136 (1935)
16. Kirkwood, J.G.: Theory of Solutions of Molecules Containing Widely Separated Charges with Special Application to Zwitterions. J. Chem. Phys. **2**, 351–361 (1934)
17. Laidler, K.J.: The development of the Arrhenius equation. J. Chem. Educ. **61**, 494–498 (1984)
18. Logan, S.R.: The origin and status of the Arrhenius equation. J. Chem. Educ. **59**, 279–281 (1982)
19. L'vov, B.V.: Activation effect in heterogeneous decomposition reactions: fact or fiction? Reac. Kinet. Mech. Cat. **111**, 415–429 (2014)
20. Marcus, R.A.: Chemical and electrochemical electron-transfer theory. Ann. Rev. Phys. Chem. **15**, 155–196 (1965)
21. Marcus, R.A.: On the theory of electron-transfer reactions. VI. Unified treatment for homogeneous and electrode reactions. J. Chem. Phys. **43**, 679–700 (1965)
22. Noyes, R.M.: Effects of diffusion rates on chemical kinetics. Prog. Reac. Kinet. **1**, 129–160 (1961)
23. Stokes, G.: On the Effect of the Internal Friction of Fluids on the Motion of Pendulums. Trans. Camb. Phil. Soc. **9**, 8–106 (1856)
24. Swain, C.G., Scott, C.B.: Quantitative correlation of relative rates. Comparison of hydroxide ion with other nucleophilic reagents toward alkyl halides, esters, epoxides and acyl halides. J. Am. Chem. Soc. **75**, 141–147 (1953)
25. von Smoluchowski, M.: Versuch einer mathematischen Theorie der Koagulationskinetik kolloider Lösungen. Z. Phys. Chem. **92**, 129–168 (1917)
26. Taft Jr., R.W.: Polar and steric substituent constants for aliphatic and o-Benzoate groups from rates of esterification and hydrolysis of esters. J. Am. Chem. Soc. **75**, 3120–3128 (1952)
27. Taft Jr., R.W.: Linear Steric Energy Relationships. J. Am. Chem. Soc. **75**, 4538–4539 (1953)

Chapter 5
Common Pitfalls

Logical traps, i.e., common and seemingly correct but in fact erroneous ways of thought, often plague scientific thinking. Some of them have achieved notable fame (or infamy), with the field of statistics providing one of the best known examples, which is known as **the prosecutor's fallacy** [21]. This logical trap is a very tempting (mis)interpretation of probabilities and provides skillful lawyers courtroom arguments that sound quite convincing but are actually quite wrong.

There are numerous such traps in chemical kinetics as well, probably more than in many other fields of science. This chapter will reveal quite a few of them and identify them as fallacies. The basic error in most of these fallacies is obvious when viewed without scientific context, but details and specific information very often cloud these issues and make the errors difficult to recognize or avoid. It is also true that virtually everyone is prone to use these flawed sequences of thought and one must learn purposefully to identify them and guard his or her thinking (and more importantly, publications) against them. The present author knows this quite well, as he recognized most of the fallacies listed here after being trapped in each of them—and not just once.

Most of the fallacies described in this chapter have been printed multiple times in scientific articles by respectable authors. This text will not cite any such specific examples for two reasons. First, it is not the intention of the author to single out any of his colleagues for their mistakes (which may have even been recognized since their publication). Second, an attempt to simply cite most of the examples would probably fill a volume on its own.

Yet, the fact that someone in the literature made a particular mistake once (or even several times) does not make the statement true, and certainly does not authorize anyone to repeat it. A further characteristic of the fallacies revealed in this chapter

© Gábor Lente 2015
G. Lente, *Deterministic Kinetics in Chemistry and Systems Biology*,
SpringerBriefs in Molecular Science, DOI 10.1007/978-3-319-15482-4_5

is that there are often attempts to picture them as opinions, whereas the error in them is clearly a matter of facts.[1] Recognizing that theories or opinions are wrong is a very natural part of scientific progress.

5.1   The Fallacy of Reaction Rates

There are two major logical blunders involving the reaction rate. The first is that it exists for any reaction. The second is that it is characteristic of a chemical reaction, or even one of its reagents.

Chapter 1 of this book defined the rate equation without using the concept of reaction rates. This is not accidental. Most reactions do not have a definite rate as they are composed of several steps, all of which have rates on their own. A careful reading of IUPAC recommendations shows that the definition of the "rate of reaction" actually says [17]: "For the general chemical reaction $(aA + bB = pP + qQ \ldots)$ occurring under constant-volume conditions, without an appreciable build-up of reaction intermediates, the rate of reaction v is defined as

$$v = -\frac{1}{a}\frac{d\,[\mathrm{A}]}{dt} = -\frac{1}{b}\frac{d\,[\mathrm{B}]}{dt} = \frac{1}{p}\frac{d\,[\mathrm{P}]}{dt} = \frac{1}{q}\frac{d\,[\mathrm{Q}]}{dt} \tag{5.1}$$

where symbols...."

Unfortunately, textbooks tend to drop the condition about the lack of an appreciable build-up of intermediates and define the rate of reaction for every process. This is quite erroneous, as the supposed equation between the rates of product buildup and reagent consumption is obviously absent when an intermediate is formed. In fact, in a system of reactions, only the individual reaction steps have rates.

The other fallacy concerning the reaction rate is that it is characteristic of the reaction or one of its reagents. In fact, the rate of reaction is almost always dependent on the concentrations of the reactants, and not only on their identity. It is not uncommon to speak about the typical time scale of a reaction, but it should at least tacitly be understood that the concentrations are in some sort of usual range that is convenient for the investigations. A very special and, from a strictly kinetic point of view, quite regrettable example of this fallacy is the widespread use of "turnover frequency" (TOF) in catalysis research. The concept and its highly questionable value in science were discussed in the current literature [3, 4, 10, 18]. Actually, the usual definition of TOF in catalytic reactions is the ratio of the rate of useful product formation and catalyst concentration. This gives a typical dimension of inverse time, which is identical to the dimension of a first order rate constant. Maybe

---

[1]To quote the late US senator Daniel Patrick Moynihan: "Everyone is entitled to his own opinion, but not his own facts."

this coincidence is the reason why it is quite often thought that a catalyst with higher TOF is more efficient. In fact, TOF is obviously dependent on the concentrations (of the substrates or even that of the catalyst) and should by no means be used to characterize the efficiency of a catalyst. Even after the specification of all reaction conditions, the rate is often calculated as a difference rather than a differential, so the measured TOF value is also dependent on the time of the measurement. The only really reliable way of comparing catalytic efficiencies is to use identical reaction conditions for all different catalyst tested. In this case, the rate of product formation provides a good basis for comparison, dividing it with the catalyst concentration to calculate a TOF value only invites intellectual sloppyness in further work.

Another phrase in widespread use is "intrinsic rate" or "intrinsic kinetics", meaning that the measured rate is free of mass transfer limitations. This specification may seem necessary in some technological applications, where mass transfer effects should indeed be considered. However, the definition of the rate of a reaction step already takes care of this issue and there is no need to specify further the lack of such mass transfer limitations. In addition, the word intrinsic has a quite different definition in IUPAC conventions in the term **intrinsic barrier** [17].

5.2 The Fallacy of Rate Constants

The fallacy of rate constants is that they are suitable for comparing reaction rates. In casual thinking, it is often inadvertently assumed that a higher rate constant will automatically indicate a higher rate of a reaction step as well.

In fact, rate constants only give reaction rates after multiplication with suitable concentrations. This remark might seem ridiculously obvious at first, but in every-day practice, it is often easy to forget in the course of making kinetic arguments.

A very spectacular example of a major blunder originating in confusing rates and rate constants is provided by the concept of "enzymatic rate acceleration". This quantity is often given as the ratio of two first order rate constants, one characterizing the studied biochemical process without the intervention of the enzyme (as the denominator), and the other one is k_2 for the Michaelis–Menten mechanism (as the enumerator). The described ratio is dimensionless and is used to illustrate the accelerating power of enzymes (sometimes the high numbers given are also meant to support statements about the extremely high catalytic efficiency of enzymes). Unfortunately, this way of thinking usually gives a highly inaccurate overall picture. In the non-enzymatic pathway, the rate constant used should be multiplied by the concentration of the substrate to obtain the rate. On the other hand, the first order rate constant of the enzyme-catalyzed pathway (i.e., k_2 for the Michaelis–Menten mechanism) should be multiplied by the concentration of the enzyme–substrate adduct to obtain the rate of reaction. This concentration of the enzyme–substrate adduct (which is, incidentally, limited by the initial enzyme concentration) is orders of magnitude lower than the concentration of the free substrate in a typical biochemical system, so the ratio of the actual product formation rates in the two pathways is a lot lower than implied by the number given as "rate acceleration."

Another example of the fallacy of the rate constant is when two (pseudo-)first order processes are compared, and the one with the larger first order rate constant is said to be faster. In fact, a first-order rate constant only measures how fast a reaction approaches its final state, but says nothing about the rate of concentration changes. A process with a lower first order rate constant may actually be faster than another one with a higher first order rate constant.

A very common mistake in interpreting reversible first order reactions is connected to the fallacy of rate constants. The scheme was given in Chap. 2 as Eq. (2.45):

$$A_1 \underset{k_2}{\overset{k_1}{\rightleftharpoons}} A_2 \tag{5.2}$$

A derivation there showed that this scheme gives rise to exponential kinetic traces with a first order rate constant of $k_{obs} = k_1 + k_2$. Without the reverse reaction, the first order rate constant would be k_1. So a fallacious interpretation is that the addition of the reverse step, paradoxically, makes the process "faster." In fact, the rates in the presence of the reverse step are never higher than in its absence, it is just the observed first order rate constant that is larger.

5.3 The Fallacy of Rate Coefficients

The fallacy of the rate coefficient is that this is the correct term to be used instead of the term rate constant. As far as this question is concerned, the diverse world of kineticists seems to be divided into two large groups. Members of the first group think that only the term "rate coefficient" is correct and also regularly attempt to correct the non-compliant usage of others. Kineticists in the second group usually prefer to say "rate constant" but do not seem to mind if others use the term "rate coefficient." Textbooks also usually follow the guidelines given by one of these two groups.

Those in favor of the term "rate coefficient" typically use two arguments to support their views against the use of the expression "rate constant." The first is that these are not constants, as certain external factors (temperature, sometimes pressure, dielectric constant, etc.) influence them. Yet, this is a misinterpretation of the mathematical meaning of the word constant. Rate constants are called constants because they are independent of the variables that appear explicitly in rate equations, which are concentrations. In mathematics, saying that "the values of the constants change in an equation" is by no means paradoxical. No sane physical chemist seems to have aversions against the phrase equilibrium constant, although its value depends exactly on the same external factors as the value of a rate constant.

The second anecdotal argument supposedly supporting the views of rate coefficient fans is that "this is what IUPAC recommendations say." In fact, even a cursory reading of the actual recommendations [8, 17] will show that this is by no means true.[2] In addition, if someone finds delight in discovering discrepancies in official recommendations, comparing the definitions of the term "rate constant" in the IUPAC glossary of terms for physical organic chemistry [17] and that for chemical kinetics, including reaction dynamics [8] will be a joyful exercise. The first text [17] uses the expressions "rate constant" and "rate coefficient" fully interchangeably. The second set of recommendations [8], on the other hand, says "It is recommended that the latter term, rate constant, be confined to reactions that are believed to be elementary reactions," and then fails to follow its own advice consistently.

5.4 The Fallacy of Consecutive Processes

The fallacy of consecutive processes is that the first process always has a higher rate constant than the second one in the series. It is quite obvious why this statement is wrong when it is put in this way. However, it is much more difficult to avoid this trap during the actual evaluation of kinetic results.

Consider two consecutive, irreversible first order reactions as given earlier in Eq. (2.57):

$$A_1 \xrightarrow{k_1} A_2 \xrightarrow{k_2} A_3 \tag{5.3}$$

Typically, some sort of instrumental reading is used to obtain the first order rate constants in this scheme and the fitted function is double exponential as described in Eq. (2.63):

$$Y_t = X_1 e^{-k_1 t} + X_2 e^{-k_2 t} + E \tag{5.4}$$

This bi-exponential curve is fully symmetric for the exchange of its two first order rate constants. Therefore, based on the values of the rate constants only, there is no way of telling which occurs first in the scheme and which is second. Instincts tell scientists that the higher (i.e., faster process) must be the first in the series, but this should in fact be proved independently on a case-by-case basis. This is possible through the careful analysis of amplitude values determined from the fits (cf. Eq. (2.59)) or by finding a way to monitor reactant A_1 selectively, which makes it possible to determine k_1 without interference from the second process.

---

[2]It is a long-standing observation of the present author that IUPAC recommendations are much more often cited than read. It is a pity because their wisdom typically exceeds the expectations of chemists by far.

Similarly to other cases discussed in this chapter, the example given here is a simple one. In real-life scenarios, the fallacy of consecutive processes often appears in forms that are made a lot more confusing by a complicated chemical background and kinetic scheme. Nevertheless, it must always be remembered that establishing the order of rate constants determined for consecutive reactions typically requires considerable efforts.

5.5 The Fallacy of Parallel Processes

The fallacy of parallel processes is that the rate constant of one of the parallel steps can be obtained by selectively monitoring its product. Even for experienced kineticists, an intuitive approach would not reveal this logical trap, which stems from the well-established chemical concept of selectivity. However, in this special case, these intuitions lead to the wrong final conclusion, which will be illustrated here using a simple example.

Consider the scheme of parallel reactions given earlier in Eq. (2.54):

$$A_1 \xrightarrow{k_1} A_2$$

$$A_1 \xrightarrow{k_2} A_3$$

(5.5)

It is very tempting to think that if A_2 is selectively detected in this reaction, then the curve describing the increase of its concentration features k_1 as the first order rate constant because of the selective monitoring. However, recalling the solution of this scheme given earlier in Eq. (2.56) shows that this is not in fact true. The solutions for $[A_2]$ and $[A_3]$ are as follows:

$$[A_2]_t = \frac{k_1}{k_1 + k_2}[A_1]_0 e^{-(k_1+k_2)t} \quad [A_3]_t = \frac{k_2}{k_1 + k_2}[A_1]_0 e^{-(k_1+k_2)t}$$

(5.6)

So both concentrations are described by an exponential function with a first order rate constant of $k_1 + k_2$, and it is impossible to determine them separately solely from the observed first order rate constant. To obtain the values of k_1 and k_2 selectively, one possibility is to determine the final concentrations of A_2 and A_3, their ratio is the same as the k_1/k_2 ratio.

On a little more philosophical level, one must note that the rate of the formation of products A_2 and A_3 actually does not depend on their concentrations, it is only $[A_1]$ that appears in the rate equation. Therefore, monitoring the products selectively does not help in separating the parallel rate constants. It should also be added that this fundamental phenomenon may appear in much more complicated schemes as well.

5.6   The Fallacy of the Rate Determining Step

The fallacy of the rate determining step is that it is slower than other steps following it. In fact, the rate of the steps following the rate determining step is usually very close to identical. Usually, a step that consumes an intermediate cannot be faster (or, at least, cannot be faster for a long time) than the rate of the step producing the intermediate, because the concentration of the intermediate would drop to very small values rapidly.

As an example, consider the scheme of two consecutive first order reactions given earlier in Eqs. (2.57) and (5.3) again with a case where $k_1 \ll k_2$ so that the first reaction in the sequence is rate determining for the formation of final product A_3. Under those conditions, the time-dependent rate of the first step (identical to the rate of the concentration change of A_1 except for a multiplication by -1) is given as follows:

$$v_1 = -\frac{[A_1]}{dt} = k_1[A_1]_0 e^{-k_1 t} \tag{5.7}$$

The rate of the second step is identical to the rate of concentration change for product A_3, but the exact formula can be greatly simplified by taking $k_1 \ll k_2$ into consideration:

$$v_2 = \frac{[A_3]}{dt} = \frac{[A_1]_0 k_1 k_2}{k_1 - k_2}(e^{-k_2 t} - e^{-k_1 t}) + k_2[A_2]_0 e^{-k_2 t} \approx k_1[A_1]_0 e^{-k_1 t} \tag{5.8}$$

A comparison of rates v_1 and v_2 shows that they are almost identical despite the fact that the first step is clearly rate determining. Among kineticists, it is quite common to say that the second process is orders of magnitude faster than the first, therefore the first is rate determining. If the word faster refers to reaction rates in the previous sentence, the statement is not technically true. However, it is also arguable that "faster" is intended to refer to the relationship of the rate constants here. For practical purposes, the minor inaccuracy in the quoted sentence is seldom an obstacle of understanding. The only really important thing that must be remembered in this context is that only rate constants with identical dimensions should compared when making decisions about rate determining steps.

5.7   The Fallacy of Exchange Reactions

The fallacy of exchange reactions is that first order kinetics (with respect to the limiting reagent) can be deduced from detecting an exponential curve. In fact, as the McKay equation (Eq. (2.43) [15, 16]) clearly shows, exchange reactions always lead to the detection of exponential kinetic traces, no matter what the rate equation of the exchange is. The key to this seemingly contradictory phenomenon is that the rate of an exchange reaction never changes throughout a single experiment, it is only the distribution of the label that changes. A minor and connected fact is that exchange reactions never have limiting reagents, as none of the reagents are consumed in the first place.

As a general guideline, the rate equation should always be determined by measuring the changes in the rate of concentration change in response to changes in concentrations. In a usual pseudo-first order curve, the concentration of one (and only one) reagent changes, so the first order dependence with respect to that single reagent can be deduced from the exponential curve shape. In the case of exchange reactions, the curve shape is exponential by default. However, the general guideline given at the beginning of this paragraph should still be followed: the rate equation can be deduced from measuring the rates of exchange in a number of experiments where different concentrations of the substances involved are used. Even in exchange reactions, kineticists tend to identify a limiting reagent intuitively. However, unlike in usual cases, the initial concentration of this limiting reagent should also be changed to obtain a full rate law.

5.8 Misconceptions About Activation Parameters

Activation is an important concept in the interpretation of chemical kinetics. Some activation parameters (typically enthalpy, entropy, and volume) are often determined for a reaction to draw conclusions for the mechanism of the process. Unfortunately, there is a very common general mistake in using activation parameters. The entire concept of activation is only valid for an elementary reaction. Therefore, activation parameters should also be reported only for elementary reactions. An overall reaction should have as many sets of activation parameters as the number of elementary reactions in it. In addition to this general remark, there are more specific misconceptions about most of the commonly used activation parameters.

5.8.1 The Fallacy of the Activation Energy

The fallacy of the activation energy is that it indicates the rate of a process. A very common way of thought is that high activation energy means low reaction rate, whereas low activation energy implies a high reaction rate. In fact, the energy of activation, as defined in the Arrhenius equation (see Eq. (4.12)), characterizes how fast the rate constant changes in response to temperature change. For calculating the rate constant (and not the rate!), a second determined parameter, the pre-exponential factor is also needed. A higher activation energy may very well be (and often is) compensated by a higher pre-exponential factor.

In fact, a high activation energy implies higher sensitivity to temperature change, whereas a lower activation energy means that the rate constant will increase less as the temperature is increased.

Interestingly, the same misconception is a lot less frequently heard in connection with the enthalpy of activation that appears in the Eyring equation, despite the fact that it is highly analogous to the activation energy.

5.8.2 The Fallacy of the Activation Free Energy

The fallacy of the activation free energy is that it carries information not already present in the rate constant. In fact, the Eyring equation (Eq. (4.14)) connects the activation free energy with the rate constant at a given temperature. Therefore, the information content of k and $\Delta^{\ddagger}G$ is exactly the same. However, there is a slight conceptual difference: a rate constant always has a well-defined meaning, whereas the activation free energy is only relevant for an elementary reaction and only within the validity range of the Eyring equation.

It seems that quoting activation free energies became widespread with the gradually intensifying reliance on theoretical calculations. In usual theoretical calculations on the kinetics of processes, energy barriers are given as a final result. So a comparison between theory (energies) and experiments (rate constants) requires some compromise. A numerical point should also be made here: the value of RT at room temperature is about 2.5 kJ/mol. In theoretical chemistry, this energy, which is usually considered as characteristic of thermal movement, is thought to be small. Most certainly, any theoretical calculation that is in agreement with experimental results within an error of RT is an excellent one in reaction kinetics (or reaction dynamics, which is the preferred termed of quantum chemists). Yet, in terms of rate constants, an error of RT in energy is a multiplication factor of about 2.7 because of the exponential function involved in the Eyring equation. Or to put it in another way, an order of a magnitude difference in rate constants (huge from an experimental point of view) is about 5.7 kJ/mol in terms of activation free energies (minor for theoreticians). This may be one of the reasons why theoretical chemists prefer comparing activation free energies rather than rate constants.

5.8.3 The Fallacy of the Activation Entropy

The fallacy of activation entropy is that its value is unreliable because it is a result of an extrapolation to infinite temperature.[3] The error in this way of thought was pointed out in a short article [9], which, quite surprisingly, is occasionally cited in the scientific literature instead of the original Eyring equation [1].

---

[3]This argument is most often used by kineticists who have access to instruments operating under very high and variable pressures. In mechanistic research, they favor activation volumes for the diagnostic purpose that activation entropies are also used for.

The Eyring equation gives the rate constant using the parameters activation enthalpy and activation entropy. At a given temperature, with a reliably known rate constant and activation enthalpy, it is possible to calculate the activation entropy without any further ado. This fact alone shows that the lack of numerical reliability is a myth.

It is understood quite well that the standard errors of activation enthalpy and activation entropy correlate quite strictly, their ratio is the average temperature of the measurements done [9]. This relation can actually be derived from the individual standard errors of the activation enthalpy and entropy, which can be calculated as given in Eq. (4.21) of Chap. 4:

$$\frac{\sigma_{\Delta^{\ddagger}H}}{\sigma_{\Delta^{\ddagger}S}} = T_{av} \left(= \frac{1}{N} \sum_{i=1}^{N} T_i \right) \tag{5.9}$$

This equation provides a nice and quick way of testing if the numerical calculations were appropriately carried out (it must be kept in mind that enthalpy is typically measured in kJ/mol, whereas the typical entropy unit is J/mol/K). If this test fails, the standard errors were determined in an incorrect way. The source of such an error is often the incorrect assumption that the relative standard error of the intercept determined from the Eyring plot is the same as the relative standard error of the natural logarithm of the activation entropy. In fact, it is the absolute standard errors that are transferable in this case.

The standard activation entropies of elementary reaction are normally between -150 and $+150$ J/mol/K. Another common mistake connected to the interpretation is stating that a $\Delta^{\ddagger}S$ value of 5 ± 10 J/mol/K is very poorly determined. In fact, the error ± 10 J/mol/K is quite typical and by no means unreliable. It must be remembered that $\Delta^{\ddagger}S$ is the difference of the entropy of the activated complex and the initial reactants. The value 5 ± 10 J/mol/K simply means that these two entropies are about equal, and meaningful scientific conclusion can be drawn from this fact.

5.8.4 The Fallacy of the Activation Volume

The fallacy of the activation volume is that it has an extremum in the transition state. In fact, the transition state is a saddle point on the energy surface, so in one direction it is a maximum, in all other directions, a minimum in energy (although it is not theoretically correct to identify this energy with enthalpy, entropy, or free energy). This property of the transition state has no consequences for activation volume at all.

Activation volumes are typically thought to be indicative of the associative and dissociative nature of ligand exchange or substitution reactions. These processes often also require some symmetry in molecular movement because of the principle microscopic reversibility, which also has some symmetry consequence for the energy surfaces characterizing these processes.

Furthermore, it is sometimes thought that activation volumes are intuitively more useful than activation entropies because the definition of entropy in thermodynamics is more abstract. Indeed, volume may be easier to imagine than entropy. However, entropy is a much more fundamental concept in thermodynamics than volume is. Entropy occurs in two of the four laws of thermodynamics, whereas volume appears in none of them.

5.8.5   The Fallacy of the Isokinetic Temperature

The fallacy of the isokinetic temperature is that it has any scientific meaning. In fact, whenever an isokinetic temperature is determined, it is just proof of a statistical correlation that has no implications for the mechanisms of reactions.

An isokinetic temperature is defined for a series of reactions with a plot of activation enthalpies against activation entropies. If the points fall onto a straight line in this plot, the series of reactions is said to be isokinetic and the slope of the straight line is called isokinetic temperature. Typically, the series of studied reactions is concluded to proceed by the same mechanism in this case. In addition, it also often said that all the reactions have identical rate constants at this isokinetic temperature.

The major error in this sequence of thought was spectacularly exposed by McBane [14]. Even for an unsuspicious scientist, it would be a galactic coincidence that more than four different reactions have identical rate constants at a given temperature. A simple look at the data themselves will reveal that this is not the case despite the linearity of the isokinetic plot. McBane also showed that the linearity of this isokinetic plot is a necessary consequence of the fact that rate constants are typically measured in a temperature range that is small compared to the absolute temperature itself [14]. In a series of reactions, a change in the rate constants by three orders of magnitude at any given temperature is usually huge. Yet this only defines a narrow band of possible values for the activation entropy–activation enthalpy combinations as shown in Fig. 5.1. In other words, the activation entropy and activation enthalpy are correlated simply because of the limited experimental range of rate constants measured. McBane even illustrated the fallacy of the isokinetic temperature by assembling a perfect isokinetic plot using the phone numbers of his friends [14].

The correlation between enthalpy and entropy is not limited to activation entropy and activation enthalpy. It is valid for the thermodynamic data as well and is often mentioned as the compensation effect in the literature [5–7, 13].

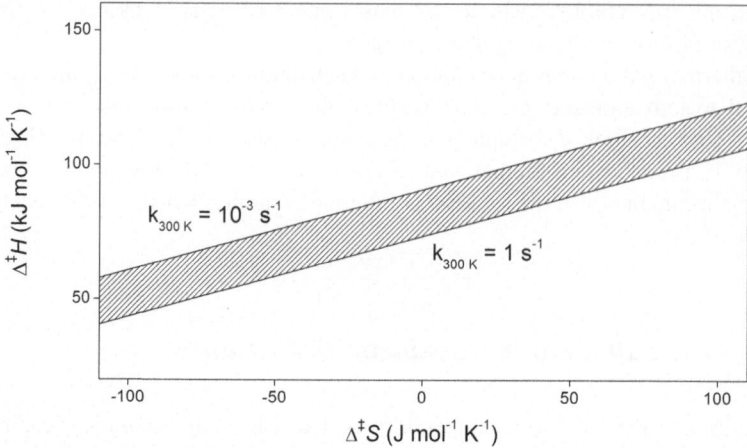

Fig. 5.1 Possible values of activation entropy and activation enthalpy for rate constants between $1\,s^{-1}$ and $10^{-3}\,s^{-1}$ at 300 K. Note that all values must fall into the narrow region between the two straight lines

5.9   The Fallacy of the Diffusion Limited Rate Constant

There are two different common lapses of logics connected to the diffusion limited rate constant. The first is that reaction rates are limited by diffusion and not rate constants. The second is that diffusion can limit first order processes.

A diffusion limited rate constant is always a second order rate constant. In this statement, both "second order" and "rate constant" should be emphasized. The commonly quoted value of the diffusion limited second order rate constant, 10^{10} $M^{-1}s^{-1}$, refers to the solvent of water at about room temperature and involves an uncertainty of almost an order of magnitude because of the approximations and generalizations used in deriving it.

A rate is not usually limited by diffusion, as it can grow as long as the concentrations can increase. The rate constant was also called specific rate some decades ago, so it is a rate normalized by concentrations. In fact, it is the rate constant (or specific rate) that is limited by diffusion and not the rate of concentration change.

The speed of diffusion may determine the speed at which two particles approach each other. In first order reactions, the collision of particles is not needed at all. Therefore, first order reactions cannot be limited by diffusion and the upper limit of 10^{10} is not valid for such reactions (in any case, this argument would also be invalidated by the mismatch of the dimensions of the rate constants). It is arguable that first order reactions may have highest possible rate constants, but that must be somehow connected to the speed of internal molecular motions as pointed in Sect. 4.1.

5.10   The Fallacy of Unimolecular Reactions

The fallacy of unimolecular reactions is that they are paradoxical.

An interpretation often repeated for bimolecular reactions is that the collision of particles having extra energy is needed to overcome the limitation posed by the activation energy in an elementary reaction. No similar line of thought can be valid for unimolecular processes because excess energy cannot be provided by a colliding particle. As a consequence, unimolecular reactions are often declared to be in conflict with the concept of activation.[4]

In fact, molecules necessarily have several internal degrees of freedom between which energy transfer is possible. For a reaction to occur, energy must be accumulated along one (or a few) of these internal degrees of freedom. The formal description of a process could even be the same as the description of the intermolecular energy transfer between different particles. Keeping these facts in mind, there is nothing paradoxical in unimolecular processes.

In a variation of this fallacy, radioactive decay is declared to occur "without any reason." In fact, atomic nuclei also have internal degrees of freedom, so an interpretation very similar to the one presented for molecules can be useful.

At this point, it should be recalled that the theoretically reasonably well-supported Eyring equation interprets a first order rate constant (i.e., one corresponding to a unimolecular process) in its original form (cf. Eq. (4.14)). In the transition state theory, it is bimolecular reactions that need extra considerations, as described in Sect. 4.2.

The fallacious argument presented here is in such a common use that textbooks often describe a scheme called the Lindemann–Hinshelwood mechanism to interpret the fact that first order reactions exist [2, 11]. In fact, this line of thought is a circular argument, as will be proved here, which does not interpret a seemingly unimolecular reaction through bimolecular activation at all.

The Lindemann–Hinshelwood mechanism posits that first order reactions actually occur because some particles gain extra energy through the interaction with other species present in the system. The scheme most often considered involves the reversible second order reaction between the main species A_1 and auxiliary species A_2 (which can be any of the surrounding molecules, even A_1 itself) to form an energetically excited form of A_1, denoted $A_1^{\ddagger}$ here, in a reversible process. No real

---

[4]This is a typical example where scientists would benefit from following an ethical guideline of the Dalai Lama: observed phenomena should not be dismissed just because of the absence of explanatory mechanisms [20].

chemical change occurs in this step. Then $A_1^{\ddagger}$ forms the product in a second step. The scheme can be represented as follows:

$$A_1 + A_2 \underset{k_2}{\overset{k_1}{\rightleftharpoons}} A_1^{\ddagger} + A_2$$

$$\text{(5.10)}$$

$$A_1^{\ddagger} \xrightarrow{k_3} A_3$$

The energetically excited $A_1^{\ddagger}$ is handled as a minor intermediate under steady state conditions. With this assumption, the rate of product formation is given as:

$$\frac{d[A_3]}{dt} = \frac{k_1 k_3 [A_1][A_2]}{k_2 [A_2] + k_3} \tag{5.11}$$

Then the explanation goes on to point out that if $k_2[A_2] \gg k_3$ holds, then Eq. (5.11) is simplified into one where the rate of product formation does not depend on the concentration of A_2 and is first order with respect to A_1. However, the condition $k_2[A_2] \gg k_3$ means that the k_3 process should be slow. This is a unimolecular step with the single reactant $A_1^{\ddagger}$. So the attempt at the theoretical interpretation of a first order process based on second order activation is still fundamentally dependent on the assumption that another unimolecular process is slow. This is a circular argument because now the unimolecular nature of product formation from $A_1^{\ddagger}$ calls for an explanation. Therefore, this author does not view the Lindemann–Hinshelwood mechanism as a scientifically valid resolution of the supposed conceptual problem with first order processes. On a more positive note, the mechanism itself can actually be useful in gas reactions to interpret cases when a reaction rate is dependent on the presence of otherwise nonreactive substances at low concentrations.

5.11   The Fallacy of Radical Scavengers

The fallacy of radical scavengers is that an assumption of a radical type mechanism can be confirmed by the fact that a radical scavenger has an effect (any sort of effect!) on the rates of concentration change.

Originally, the idea of using radical scavengers was that they react with certain, previously encountered radicals quite rapidly, and the product formed in this process (typically some sort of adduct) is easily detectable. Over time, this concept underwent a complete overhaul: if the addition of any supposed "radical scavenger" has any sort of effect on the time dependence of the concentrations, the reaction is concluded to involve radicals. Needless to say, this approach is completely wrong. Any substance can react (at least potentially) with radicals as the essence of being a radical is high chemical reactivity. So basically, everything could be considered a radical scavenger. In addition, a substance can exert an effect on a reaction rate in a number of different ways, only one of which is scavenging the radicals formed. It is much better to stick to the original concept: the use of a radical scavenger is only conclusive if it is attempted for a specific radical and the result is demonstrated by the detection of the characteristic adduct.

On a more general note, no direct conclusion can be drawn for the mechanisms of a reaction from the simple fact that a certain substance influences the reaction rate. Such an influence must usually be quantified and specific information must be obtained about the reaction step that involves the added new substance.

5.12   The Fallacy of Linearization

The fallacy of linearization is that it should be used whenever possible. In fact, for a statistically favorable data evaluation, nonlinear fitting of untransformed equations is clearly desirable.

Linearization of nonlinear functions for the purpose of fitting was conceived out of necessity in an age when the mathematical principles of data analysis were already very clear, yet the computational tools to do the right thing were missing. At that time, the best common fitting tool was a ruler, which is linear. Therefore, the only really convenient way of fitting was mathematical transformation of the original equation into a linear form, then preparing the linearized plot and use the ruler to the best judgment of the investigator to find the best fitting straight line.

Since this age, one of the major driving forces in developing computers has been to provide a way to carry out more and more detailed and statistically more and more acceptable scientific calculations. When personal computers became common, linearization methods have become obsolete. In today's science, preparing a linearized plot with a computer is akin to using a state-of-the-art oscilloscope as a tool to drive in a nail. The investigator does not even have to be aware of how such mathematical fitting works in detail: all it takes to become familiar with the use of a suitable scientific fitting software.[5]

It is no small irony, and reflects quite negatively on the usefulness of the number of citations as an indicator of scientific value, that the linearized form of the Michaelis–Menten equation (Eq. (2.12)), which is called the "Lineweaver–Burke plot" today [12], was first published in a paper that became the most cited article of all times from the **Journal of the American Chemical Society** [19]. This linearization is an arithmetic transformation that science majors at university were supposed to perform routinely even at the time when this seemingly seminal paper was published. In addition, the linearization involves inverting both axes (concentration and reaction rate). The Michaelis–Menten equation describes a

---

[5]Just to make sure: the commonly used software Microsoft Excel is suitable for a lot of tasks, but not for routinely fitting nonlinear curves.

saturation curve, which reaches a final region where the rate is not dependent on the substrate concentration any more. One of the two parameters in this equation is the final value reached. However, in the double inversion linearization credited to Lineweaver and Burke [12], this important region is transformed into a single point, which is by no means favorable for determining the maximum rate parameter.

To give another example, the linearized form of the exponential curve shown in Eq. (2.10) is still in very widespread use. This is conceptually wrong, as the curve has three parameters, so one of them has to be estimated prior to the linear fitting. This prior estimation is often done for the endpoint, so $\ln(Y_t - E)$ is typically plotted as a function of t in the linearized plot. In such a plot, the evaluation cannot usually be done to high conversions because even very small experimental errors at the end of the curve are greatly magnified by the plot type used. So in this linearization, data at the end of the curve are typically ignored despite the fact that they are not any less reliable than the points measured earlier.

Despite the arguments presented in this section, linearization is still a (somewhat) acceptable option in data evaluation in a few exceptional cases. Such cases are the Arrhenius equation in Eq. (4.12) and the Eyring equation Eq. (4.14), as explained in Sect. 4.2.

5.13   The Fallacy of the Coefficient of Determination

The fallacy of the coefficient of determination is that it characterizes the goodness of a fit. The regression coefficient is also sometimes called R^2 value, or, somewhat imprecisely, regression coefficient. Any software that can fit a straight line is also able to calculate this value. Many researchers use this number to characterize the general goodness of a fit and select some arbitrary cut-off value for decisions of accepting or rejecting a fit, usually one that fits the actual purpose of the investigator.

The error of this sort of thinking is very easily given graphically, as illustrated in Fig. 5.2.

The two data series given in Fig. 5.2 illustrate that a low number of points with reasonable random scatter may give rise to an R^2 value that is less favorable than that for a higher number of points with errors giving a very easily recognized tendency. The catch here is that the calculation of R^2 values involves a division by the number of points used for the fitting at some point, so a higher number of considered points is almost automatically translated into a more favorable R^2 value.

In fact, the regression coefficient R^2 is just one of the statistical descriptors that can be used to characterize the results of a fitting procedure. There are a number of further descriptors as well, but none of them have universal value in science. In the example shown in Fig. 5.2, calculating the serial correlation would probably give away the difference that is obvious at first sight in the graph.

In a broader context, the rigorous use of statistics is not without a catch in chemical kinetics, either. When testing assumptions (i.e., whether a curve fits to the experimental data), a choice should always be made about the level of significance

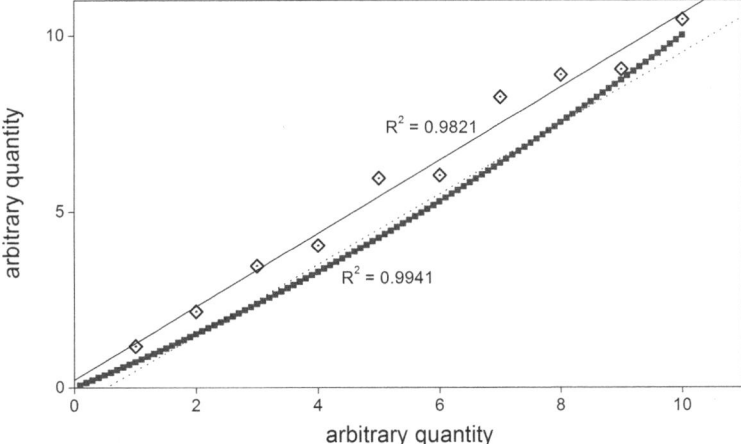

Fig. 5.2 Demonstration of the pitfall of using regression coefficients to characterize the goodness of a fit. Larger markers fit the straight line (*solid*) reasonably with some random scatter, whereas the smaller markers are clearly not described adequately by the fitted straight line (*dotted*), despite the better R^2 value

required. This level of significance cannot and should not be determined by the quality of the fit itself, it must always be based on external information.

In statistics, two different kinds of errors may be made in such assumption testing (whether a fit is acceptable overall or not) on an experimental data set. The first error is accepting an assumption when it is in fact not true, the second is rejecting it when it is in fact true. Choosing a significance level only chooses which of these two errors is more tolerable to the investigator. Lowering the chance of one kind of error will automatically increase the chances of the other. The decisions about the appropriate levels of significance are often made in other branches of science considering the balance of costs (material or other) involved in the two different kinds of mistakes. Approving an unsafe medicine seems much more unacceptable to humans than rejecting the application of an otherwise useful drug by mistake. Therefore, high significance levels (typically standardized by government authorities) are required in the pharmaceutical industry in such tests. There are no similar standardized sets of significance levels in scientific data evaluation. This is just as well, as there is also no scientific reason for seeking such generally agreed-upon guidelines. Data evaluation depends on a lot of factors and it is the primary objectives that define the criteria for acceptability.

No matter how much the investigator might seek objectivity in evaluating data, at the end of the day, there will be a point where subjective decisions must be made. Another danger of using advanced statistical methods is that these calculations may become an end instead of a means to achieve a scientific goal. Furthermore, the mathematically proficient investigator should regularly remind himself or herself that no amount of statistical sophistication can overrule good common sense.

5.14   The Fallacy of Curve Fitting

There are two fallacies about curve fitting, which, interestingly, contradict each other. The first is that the purpose of curve fitting is to learn the values of certain parameters in a theoretical function. The second and equally incorrect misconception is that curve fitting is needed to find a function that describes measured data within some predetermined precision.

The primary purpose of curve fitting is model validation: it must be demonstrated that the chosen theoretical function provides an acceptable interpretation of the measured data. The model is validated if the fitted curve and the experimental points show reasonable agreement (although what counts as reasonable may actually depend on the problem). During the evaluation of curve fitting results, the user must first decide whether the fit is acceptable. If it is not, then the parameters obtained as a result of the fit are absolutely meaningless. Information is only carried by the parameters if the acceptability test is passed first. If the assumed theoretical function does not interpret the data well, its parameters cannot be used to draw scientific conclusions.

The other misconception emphasizes the predictive value of curve fitting. Simply predicting the numerical results of experiments is sometimes a worthwhile scientific or technological objective. If this is the case, then finding a good function for this prediction is a respectable goal. However, scientific research is typically about interpreting experimental findings. In chemical kinetics, the ultimate goal is to find the series of elementary reactions that is in agreement with the detected kinetic traces. The physically meaningful interpretation typically limits the possible range of theoretical functions greatly. Therefore, the main goal of curve fitting is mostly about finding a physically meaningful theoretical function and determining the parameters in it.

References

1. Eyring, H.: The activated complex in chemical reactions. J. Chem. Phys. **3**, 107–114 (1935)
2. Hinshelwood, C.N.: Some observations on present day chemical kinetics. J. Chem. Soc. 694–701 (1947)
3. Kozuch, S., Martin, J.M.L.: "Turning Over" definitions in catalytic cycles. ACS Catal. **2**, 2787–2794 (2012)
4. Kozuch, S.: Reply to comment on "Turning Over" definitions in catalytic cycles. ACS Catal. **3**, 380–380 (2013)
5. Krug, R.R., Hunter, W.G., Grieger, A.A.: Enthalpy–entropy compensation. 1. Some fundamental statistical problems associated with the analysis of van't Hoff and Arrhenius data. J. Phys. Chem. **80**, 2335–2341 (1976)
6. Krug, R.R., Hunter, W.G., Grieger, A.A.: Enthalpy–entropy compensation. 2. Separation of the chemical from the statistical effect. J. Phys. Chem. **80**, 2341–2351 (1976)
7. Krug, R.R., Hunter, W.G., Grieger, A.A.: Enthalpy–entropy compensation: an example of the misuse of least squares and correlation analysis. Chemometrics: Theory Appl. (ACS Symp. Ser.), Chapter 10 **52**, 192–218 (1972)

8. Laidler, K.J.: Glossary of terms used in chemical kinetics, including reaction dynamics. Pure Appl. Chem. **68**, 149–192 (1996)
9. Lente, G., Fábián, I., Poe, A.: A common misconception about the Eyring equation. New J. Chem. **29**, 759–760 (2005)
10. Lente, G.: Comment on "Turning Over" definitions in catalytic cycles. ACS Catal. **3**, 381–382 (2013)
11. Lindemann, S.R.: Note on the theory of the velocity of chemical reaction. Phil. Mag. **40**, 671–674 (1920)
12. Lineweaver, H., Burk, D.: The determination of enzyme dissociation constants. J. Am. Chem. Soc. **56**, 658–666 (1934)
13. Liu, L., Guo, Q.X.: Isokinetic relationship, isoequilibrium relationship, and enthalpy–entropy compensation. Chem. Rev. **101**, 673–696 (2001)
14. McBane, G.C.: Chemistry from telephone numbers: the false isokinetic relationship. J. Chem. Ed. **75**, 919–922 (1998)
15. McKay, H.A.C.: Kinetics of exchange reactions. Nature **142**, 997–998 (1938)
16. McKay, H.A.C.: Kinetics of some exchange reactions of the type RI + I*- RI* + I- in alcoholic solution. J. Am. Chem. Soc. **65**, 702–706 (1943)
17. Muller, P.: Glossary of terms used in physical organic chemistry. Pure Appl. Chem. **66**, 1077–1184 (1994)
18. Ritter, S.: The turnover fallacy. Chem. Eng. News **91**(9), 46–47 (2013)
19. Stang, P.J.: 124 years of publishing original and primary chemical research: 135,149 publications, 573,453 pages, and a century of excellence. J. Am. Chem. Soc. **125**, 1–8 (2003)
20. Sternberg, E.: A compassionate universe? Science **311**, 611–612 (2006)
21. Thompson, W.C., Schumann, E.L.: Interpretation of statistical evidence in criminal trials: the Prosecutor's fallacy and the defense Attorney's fallacy. Law Hum. Behav. **11**, 167–187 (1987)

Index

© Gábor Lente 2015
G. Lente, *Deterministic Kinetics in Chemistry and Systems Biology*,
SpringerBriefs in Molecular Science, DOI 10.1007/978-3-319-15482-4